Coming to Life

Coming to Life

HOW GENES DRIVE DEVELOPMENT

CHRISTIANE NÜSSLEIN-VOLHARD

Yale University Press
New Haven and London

First published in Germany by C. H. Beck, 2004

First published in the USA by Kales Press, 2006

This edition published by Yale University Press, 2006

Helga Schier, Ph.D., Translator from the original German
Christopher Middleton, Translator (passage from Goethe's *Metamorphosis of Animals*)

The publication of this work was supported by a grant from the Goethe-Institut

For information about this and other Yale University Press publications, please contact:
U.S. Office: sales.press@yale.edu www.yalebooks.com
Europe Office: sales@yaleup.co.uk www.yalebooks.co.uk

Library of Congress Control Number: 2006922006

ISBN 0-300-12080-X

A catalogue record for this book is available from the British Library

10 9 8 7 6 5 4 3 2 1

Printed in the United States of America

Every animal is an end in itself, it issues
Perfect from Nature's womb and its offspring are equally perfect.
All its organs are formed according to laws that are timeless,
Even a form very rare will hold to its type, though in secret.

—Johann Wolfgang von Goethe
Metamorphosis of Animals, 1806

CONTENTS

PREFACE

THERE IS NOTHING MORE FASCINATING THAN LIFE. IT SEEMS A MIRACLE. FROM an egg—that is little more than a yolk sac protected by a thin shell—a walking, seeing, feeding chick develops. All it takes is a little time and the proper temperature. All living beings, including that chick and ourselves, are made up of "nothing" more than molecules, molecules consisting of atoms of different elements, most notably carbon, oxygen, nitrogen, and hydrogen. The structure of these molecules determines the way they react. The laws of physics and chemistry tell us that only chaos will develop on its own. And yet, in living beings order does prevail. So the question is: how does order prevail and, more importantly, how does order evolve in the first place?

Living beings are able to multiply themselves. This means that the organisms themselves are the blueprints for future generations. The nucleic acids that make up the genes provide the recipe for their own doubling. However, genes cannot replicate themselves on their own. Viruses, for example, are genes encased in protein. Yet they cannot be described as living things as they can only multiply in the context of a cell. Hence, cells, not genes, are the smallest units of life. Cells contain genes, but also harbor the building blocks and enzymes needed to multiply them. The genes provide the cell with a set of instructions that allow it to grow, to divide, to differentiate, and eventually, to build a new organism.

This book deals with the embryos of animals and their genes. How is it that an animal can develop from a fertilized egg that bore no resemblance whatsoever to it? What exactly is contained inside that fertilized chicken egg other than the nutrients providing the energy that, as the laws of physics and chemistry explain, are indispensable for the creation and preservation of order? What are the processes that lead to the development of cells of different types? And what is their characteristic succession that allows these processes to build a functioning animal that can feed itself? And finally, why do offspring look like their parents?

In the past 50 years, great strides have been made with regard to the understanding of these processes. The elucidation of the structure and function of DNA has made it possible to decipher genes and to understand the proteins they produce. The new science of molecular biology allowed not only the isolation of genes but also the visualization of their activity within a living organism. This development has brought us closer to solutions of problems that had previously been inaccessible. Answers to the mystery of how life develops now seem within closer reach than previously imaginable.

When I was a student, books of developmental biology were rather voluminous, full of highly complicated experiments pointing out so many "trees" that it was impossible "to see the forest." Several contradictory theories and opinions stood against one another, and it was difficult to know what was important, what you needed to learn and remember, and what was prudent to forget. Today things are rather different as we have a widely accepted concept of genetics and development as a common denominator.

I have written this book for all those who are curious and who would like to understand the processes of life a little better without having to deal with highly specialized knowledge. It is for those who encounter biology in their professional life: medical professionals, chemists, and physicists. I hope my readership will also include those who might have many questions regarding genes and embryos owing to current bio-political debates, such as philosophers, lawyers, politicians, and theologians. Finally, while this work is by no means a course textbook, I believe that it may help teachers and students to fill the gaps and find connections between various disciplines.

My intention is to give a brief and concise overview without getting mired in the details. But the specialized terminology cannot be avoided altogether because the structures and molecules of organisms are not exactly the topics of everyday conversation. The glossary explains the more frequently used terms. The text assumes a basic knowledge of biochemistry and presents the concepts of cell biology rather briefly. We are not dealing with the attributes and characteristics of molecules so much as with their role in the development and preservation of complex shapes and forms. This very complexity creates the richness of all living things.

Explaining the mechanisms that lead to the development of patterns and forms is one of the main concerns of this book. Embryos develop through cell and molecular interaction in time and space. It takes patience to follow as more and more components get involved. The explanations of these processes have been simplified as much as possible and therefore may appear much simpler than they are in reality. Still, in order to enjoy a basic understanding of these processes not every detail is necessary. Broad strokes will have to do because life is complicated.

The text is arranged in chapters that are to some degree autonomous. The first chapter deals with the theory of evolution and provides a vital basis for an understanding of the larger context. This is followed by an introduction to the basics of cell and molecular biology. The main body of the book deals more specifically with the questions of how genes drive development and how form arises. Here I refer mainly to the fruit fly *Drosophila* as a test subject, as it, more than any other animal, has allowed for a relatively easy glimpse into the seemingly mysterious events of pattern formation. A further chapter reviews the most important processes of the growth and development of shapes and form on the cellular level. While portions relating to the development of vertebrates may be somewhat complex for the lay reader, this complexity is necessary for understanding the development of the human embryo. The final chapters discuss relevant topics of human biology and evolution, genome research, and bio-political questions. I hope the illustrations make these seemingly complex processes a bit more understandable.

For those of you who, after reading this book, would like to know more, I suggest referring to textbooks on the subject. Today's standard work is *Molecular Biology of the Cell*, compiled by Bruce Alberts and his colleagues. Weighing in at 1,500 pages, it is much more voluminous than this work, and yet still does not manage to cover everything. For texts that are more focused on embryonic development, refer to the books by Lewis Wolpert, Scott Gilbert, and Jonathan Slack. Also, John Gerhart and Marc Kirschner describe insightful aspects of the evolution of developmental mechanisms in *Cells, Embryos and Evolution*. Helpful illustrations of the anatomy of development can be found in Ulrich Drews' *Atlas of Embryology*.

Finally, I would like to thank many friends and colleagues who have read, reviewed, critiqued, and encouraged this work at all stages. I have received many clarifications and suggestions that have served to improve my manuscript. Therefore, my sincerest thanks go to Christopher Antos, the late José Campos-Ortega, Darren Gilmour, Matthias Hammerschmidt, Dietrich Klose, Jana Krauss, Maria Leptin, Florian Maderspacher, Georg Otto, Siegfried Roth, Maria and Harald Schnabel, Frank Schnorrer, Anne Spang, Detlef Weigel, and Ernst Ludwig Winnacker. I also wish to thank Kenneth Kales for his patience and care in preparing the English version of my book for publication, and am indebted to Florian Maderspacher for his help with the translation.

Christiane Nüsslein-Volhard
Tübingen, Germany
March 2006

CHAPTER I

Origin and Heredity

Bears couple in winter. After coupling, the female withdraws into the cave and 30 days later gives birth, usually to five cubs. Bears, when first born, are shapeless masses of white flesh a little larger than mice, their claws alone being prominent. The mother then licks them gradually into proper shape.

Pliny, the Elder (23–79 AD)

THE QUESTION OF HOW AN ANIMAL DEVELOPS HAS PREOCCUPIED HUMANS since Antiquity. As long as the development of animals and humans took place in secrecy, that is, before the invention of the microscope, the origin of animals was a matter of theoretical debate rather than scientific proof. One of the more popular ideas relating to human development was the hypothesis of preformation or encasement. This theory assumed that the growing organism had already been entirely preformed in the sperm, as a so-called homunculus, and would subsequently unfold within the mother, much like a planted seed does in a flower bed.

Later theories would transfer the preformation into the egg, postulating that the egg already contained the form of the growing organism in its full complexity even though it was invisible. According to this theory, the presently growing organisms already contain the next generation that, in turn, contains the next, and so forth and so on. This would mean that the first human would have had to contain all present and future generations, like a Russian doll, a scenario that is clearly impossible. Therefore, preformation was ultimately an unsatisfactory explanation because there must be mechanisms that allow complex forms to develop anew in every single generation.

Nevertheless, preformation was the dominant theory up to the mid-nineteenth century, in part because it was almost impossible to imagine any other explanation for the phenomenon of life. At that time, people also believed in so-called spontaneous generation, for instance, that mice developed spontaneously out of rags and ducks grew on trees!

During the course of embryonic development, complex animal forms develop from simpler ones. An individual organism begins its life as a single, fertilized egg cell bearing no resemblance whatsoever to the eventual shape and organization of the fully grown animal. Often development occurs in a stepwise manner and the juvenile larval stages are transformed through more or less complete metamorphosis into the mature adult.

This phenomenon presented a multifold puzzle: how does something complex develop from something simple? And equally intriguing, why does an organism that develops from an egg ends up looking just like its parents? Why does a maggot always produce a fly? Why does a chicken egg always produce a chick and not an alligator? These questions hint at a close relationship between individual development and heredity, the passing down of specific characteristics and attributes from one generation to the next, and a link between development and genetics, embryos, and genes.

1. The Natural System: Carl Linnaeus

Aristotle (384–322 BC) is the founder of biology, the study of life. He observed and described, collected, and connected facts and was the first to formulate a general theory of life. Aristotle already distinguished between the notion of preformation and the concept of new patterns arising through each generation. He favored the latter. He described the development of the chick inside the egg and concluded that complex forms develop from simpler ones. Of course, some of his assumptions proved to be wrong; as it turns out, eels do not develop from worms. Even so, Aristotle's greatest contribution to biology was the description and classification of organisms based on their external characteristics—birds, mammals, insects, and spiders—rather than on criteria of lifestyle, such as animals living on land versus those living in water. His classifications made much sense and provided insight into the interrelationship of organisms.

The need to bring order to the ever-expanding collection of plant and animal descriptions remained unaddressed until finally the Swedish naturalist Carl Linnaeus (1707–1778) made a massive leap. He founded the basis of the classification that is still in use today: a "natural" system (*Systema naturae*, 1735) that could be applied to both animal and plant species. He based this system on a multitude of attributes common to different species. A species is a group of animals or plants that recognize each other, accept each other as reproductive partners, and can reproduce with each other.

Linnaeus grouped similar species into genera and named the species with double names such that the first part of the name refers to the genus. For example, *Parus* refers to the genus tit and *Parus caerulaeus* to the species bluetit. Similar genera are categorized into larger groupings. The largest groups are called phyla, such as mollusks that include mussels and snails, arthropods that include insects, crabs, and spiders, and chordates that include vertebrates like frogs, fish, snakes, birds, and mammals. This system has been modified and adjusted over time, incorporating new discoveries and observations, for example, when two formerly separate species are discovered to merely be the larval and adult stages of one and the same species.

In 1828, the German-Estonian biologist Karl Ernst von Baer (1792–1876) was the first to recognize that there are more similarities between the embryos of different species than there are between their adult counterparts. It is indeed striking that the embryo of a fish looks much more like the embryo of a bird than an adult fish looks like an adult bird. But until the discoveries of British naturalist Charles Darwin (1809–1882), these similarities were merely seen as a lack of creativity on the part of the Creator. Darwin recognized though that these similarities are signs of a true biological relationship, pointing toward a common origin, and thus, represent one of the most important pillars of his theory of evolution.

2. *The Theory of Evolution: Charles Darwin*

During Charles Darwin's time, it was known that the shape of the earth was changeable. Mountain ranges grew, landmasses sank, and climactic upheavals occurred throughout history; each one bringing about a change in the conditions of life on the planet.

Darwin realized that species are changeable too. There were species that had become extinct, many of which he had inspected as fossils on his expeditions around the world. There were also species that had not existed for very long, such as the finches on the Galapagos Islands—they were different on every isle, yet clearly related as if they had a common ancestor that originally populated the islands. He attempted to explain this phenomenon, creating a theory from his observations based on natural variations and the preferred reproduction of the fittest specimens. Even though the origin of a new species is very slow—taking thousands, often millions of years and cannot be directly observed—Charles Darwin was able to use numerous convincing observations to support his theory, which he published in 1859 in a book entitled *The Origin of Species by Means of Natural Selection*.

Surplus. In theory, two offspring per one pair of parents is enough to preserve the species. Yet, usually many more are born in order to compensate for offspring that will probably perish before reaching fertility. Juvenile stages are often vulnerable, and the animal may have to overcome illness and other life-threatening circumstances. The degree of "wastefulness" varies between species. Female fish that do not tend to their brood lay thousands of eggs, whereas an elephant has only few young but cares for them very well. In both cases, if none of the offspring die before they reproduce, the population will increase in a manner that is proportional to the current number of animals. This explosion of the population will eventually lead to a catastrophe once all resources have been exhausted. However, by and large, the population increase is slowed because some individuals will fall prey to predators, cold weather, competition, drought, or stupidity.

Variation. Darwin also observed that a set of parents' offspring was not identical, but instead small, random variations could be observed among the individuals of a species. Random variation is an important factor for the survival of a species, because it fosters those individuals who are better adapted to survive, should the conditions of life change. If temperatures drop, animals with more fur will have the advantage. If temperatures rise, those with less fur will be better off. If animals are exposed to a massive bacterial infection, some will prove to be resistant and will survive unscathed. Importantly, he postulated that many such variations are hereditary. This means that also the descendants of the well-adapted animal will exhibit these favorable qualities. He also noted that asexual reproduction resulted in fewer variations. But when males and females are involved, as is the case with most species, new combinations develop, increasing the occurrence of variations. Variation also affects the characteristics of sexual attractiveness to potential partners, thereby increasing the likelihood of numerous offspring. Therefore, the process of evolution is driven not only by the selection of the big and strong, but also by the selection of the attractive. Those considered beautiful, those with an appealing smell, or a sound pleasing to other members of the species have an advantage. The theories of surplus and hereditary variation are concepts that make sense and are no longer in dispute.

Selection. Charles Darwin hypothesized that, of any given offspring, those that are better equipped to withstand the challenges of life and their surroundings will have a higher chance to survive. Therefore, they will have more offspring than their less-adapted siblings, leading to "the preservation of favored races (or subspecies) in the struggle for life." This process of "natural selection" happens in small steps over an extremely long time. However, it does accelerate when groups are isolated, such as in niches or on islands, and can no longer mingle with less-adapted specimens. Darwin illustrated the principle of selection using the breeding of domestic animals as an example. Every dog breeder knows that in any given litter not all pups are the same. The breeder will, of course, choose those pups he finds most valuable for further breeding. Even during the breeding process, the variations are random and hereditary. Hereditary variations include, for example, the short legs of a dachshund. But the cropped ears on the same dachshund are not a result of heredity, but rather an example of an acquired characteristic, which cannot be passed down to the next generation. "Artificial selection" leads to divergent races of domestic animals with qualities so radically different from the original species as to be considered a new species altogether. Thus, unlike natural selection that takes an extremely long time, artificial selection can produce large differences very rapidly.

The theory of evolution—that is, the transformation of species through natural selection—was in opposition to the then popular ideas of spontaneous generation and creation. Even though the ability to classify plants and animals according to a natural system already pointed toward a common ancestor, and thus toward evolution, researchers preceding Charles Darwin assumed that similarities among differing species were a result of the Creator's plan. However, even before Charles Darwin, the German poet Johann Wolfgang von Goethe (1749–1832) did discern that animals were related to each other, and humans, therefore, ought to be categorized on one level along with other mammals.

Life came into being more than 3 billion years ago, first in the form of very simple, single-celled organisms. Multicellular organisms in which cells differentiate in order to fulfill various functions appeared much later, about 600 million years ago. The animals known to us today most likely go back to one common multicellular ancestral form.

The origin and relationship of organisms to one another are often illustrated by a branching tree. So, today's living animals would be shown on the upper branches and extinct species would be represented on the lower branches. More than 99% of all species that ever existed are extinct today. Certain species may change dramatically through the course of evolution and new species may develop via the splitting of one of the branches. Yet other species may remain unchanged over millions of years. This description leads to the assumption that most of today's species cannot trace themselves to a species existing today, but rather to an extinct ancestor species bearing some resemblance to the ones living today. For example, humans do not descend from apes, but humans and apes have a common ancestor.

The similarities and differences of embryonic forms compared to their mature forms were one of the most important pillars of Charles Darwin's theory of evolution. He argued that if creation had actually taken place, the development of an animal from its juvenile to its mature shape would be quite straightforward. In other words, there would be no reason for the larvae of snails and worms to be so similar, and no reason that mammalian embryos would exhibit gill slits that only fully develop into gills in fish. His explanation for these phenomena was that variation and adaptation of the adult animal, along with the number of its children, will determine which direction evolution will take. In contrast, embryonic shapes and forms are much less relevant for selection and thus are much more stable throughout evolution.

Darwin's theory on the origin of species by variation and selection, already conceived around 1840 but published only in 1859, greatly disturbed his contemporaries as it questioned the idea of man as Creation's crowning glory. The fact that variation and selection are vital principles that play a major role in many contexts is beyond dispute. Charles Darwin himself never attempted to explain social phenomena with his theory, but confined his thoughts to the organic world.

The important conclusion of this theory is that the characteristics of living beings are subject to the laws of evolution. All currently existing beings originated from forms that have survived during the process of evolution. They are not the result of the unfathomable design of a Creator, but have developed as a result of biological mechanisms that have been tested and improved upon. Evolution is accidental. Its driving force is the process of selection rather than the goal-oriented adaptation that might result from characteristics favored throughout an individual's lifetime. Charles Darwin's theory has been consistently supported by modern biological research, and evolution as an explanation of the origin of new species can no longer be objectively or intellectually disputed.

3. The Laws of Heredity: Gregor Mendel

As Charles Darwin was unaware of the mechanisms of heredity, he did not know how the observed variations among individuals of a species could be explained. He himself still considered inheritance of acquired features as a means promoting the appearance of new species. Genes had yet to be discovered.

The principles according to which features are passed on from one generation to the next were eventually revealed through the crossbreeding of spontaneous variations within a species. Such so-called mutations play a large role in the breeding of animals and plants. The Bohemian Augustan monk Gregor Mendel (1822–1884) examined peas that differed from each other in several characteristics: their flowers were either red or white and their seeds either yellow or green. He then crossed red-flowered peas with white-flowered peas, repeating the crossbreeding with their offspring over several generations. He also counted exactly how many of the seeds gave rise to white-flowered peas as opposed to red-flowered peas. These counts resulted in strange numerical ratios that he explained as follows.

Hereditary factors in both animals and plants appear in duplicate with one from the mother and one from the father, which he considered to be of equal significance. Gregor Mendel was the first to postulate that these hereditary factors were discrete units passed on from one generation to the next, undivided and independent of each other. These hereditary factors were later called genes.

Red and white flowers are characteristics that are created by alternative states of a single gene. Such states are called alleles. Alleles are either dominant, indicated by upper case letters, or recessive, indicated by lower case letters. In flowering peas, the red allele is dominant while the white allele is recessive. Therefore, if at least one of the two alleles in the plant is the red allele, the flower will be red. In other words, both *AA* and *Aa* will result in red flowers. Likewise, all white flowers will be *aa*. If both copies of the gene are identical, the organism is called homozygous, because all offspring will look like the parents ($aa \times aa \Rightarrow aa$). If the alleles are different, that is *Aa*, the organism is called heterozygous.

When homozygous individuals with different characteristics are crossbred, the descendants will all look the same, each exhibiting the dominant characteristics ($AA \times aa \Rightarrow Aa$). This is because homozygous parents will create only one kind of germ cell or gamete. More specifically, *AA* individuals will create *A* germ cells, and *aa* individuals will create *a* germ cells. Therefore, only one combination is possible in the first filial generation or F1, namely, *Aa* (Figure 1, The Law of Uniformity).

When such heterozygous F1 individuals are crossbred, the recessive characteristic will occur again in the next generation F2 but only in one quarter of the descendants. For example, red peas versus white peas will appear in a ratio of 3 to 1 since there are now two kinds of gametes possible, *A* or *a*. The resulting offspring will then be of one of the four different possible combinations: $A + A$, $A + a$, $a + A$, or $a + a$ (Figure 1, left panel, The Law of Segregation).

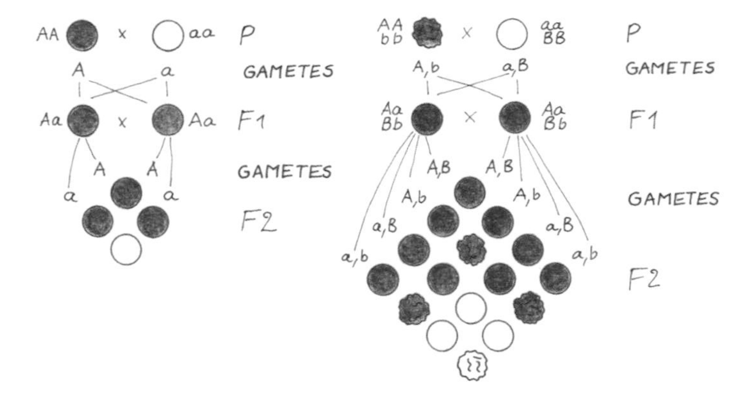

Figure 1. *Mendel's Laws Explained on Peas.* The left panel shows the pattern of inheritance for one trait (color of the flower) and the right panel for two traits (color and smoothness of the pea). Gametes are germ cells; in plants they are egg cells and pollen and in animals they are eggs and sperm. P: parental generation; F1: first filial generation – the children. F2 are the grandchildren. *AA* = red flower, *aa* = white flower, *BB* = smooth pea, *bb* = wrinkled pea. The constitution of the filial generations comes from the independent combination of gametes.

What is true for A also holds for B. Indeed, several different characteristics can be passed on independently at the same time. Here too, following the Law of Uniformity, parents that are homozygous for two independent traits will create only one kind of offspring, F1 individuals that are doubly heterozygous. When crossbreeding such *Aa* and *Bb* individuals with each other, dominant and recessive characteristics of both traits will appear in the ratio of 3 to 1. Such crossbreeding will also create a combination that neither appeared in the parent nor in the grandparent generations, namely descendants that show both recessive characteristics of *aa* and *bb*. Only 1 out of 16 descendants will exhibit the characteristics of both recessive genes (Figure 1, right panel, The Law of Independent Assortment).

According to Mendels theory, the germ cells or gametes, in the case of animals the egg cells and sperm cells, carry only one copy of every gene, whereas body cells (somatic cells) carry two copies. The genetic constitution of the somatic cells determines the phenotype, that is, the trait that is actually apparent.

In summary, Gregor Mendel's extraordinarily brilliant conclusions were that hereditary factors are discrete, indivisible, and independent units passed on through the germ cells from one generation to the next. He postulated his rules at a time when chromosomes had not yet even been discovered and the processes of fertilization were still a mystery. He realized that his rules of heredity were only relevant if they held true for other organisms as well. But it took many years to support his hypotheses in other plants and animals. This is one of the reasons why his work, first published in 1866, was overlooked until it was rediscovered in 1900.

CHAPTER II

Cells and Chromosomes

IN THE NINETEENTH CENTURY, THE MAIN TOPICS OF ZOOLOGY DEALT WITH the discovery, description, and classification of new species. Embryology, the study of the development of animals, reached its heyday at the end of the nineteenth century. Many centers for marine biology, such as those in Naples, Italy or in Woods Hole, Massachusetts, were founded at around this time. The eggs of numerous marine animals were readily available and the embryonic development of the often translucent specimen could be studied in living samples, or with easy methods of fixation and staining. In 1827, German-Estonian biologist Karl Ernst von Baer (1792–1876) described the first human egg cell that was derived from a young woman who had drowned after a wild night. However, as mammalian eggs and embryos were not easy to come by, the study of mammalian development remained on the back burner for a long time. By contrast, egg-laying animals, like sea urchins, frogs, fish, and worms, proved to be excellent objects for the study of the early stages of embryogenesis.

With the help of the microscope it was discovered that organisms consist of cells that divide, and that embryos develop from simple to more complex forms. Even in the mid-nineteenth century people still seriously discussed spontaneous generation, the formation of living beings from dead matter, until the German medical researcher Robert Remak (1815–1865) finally showed that every cell springs from a precursor cell. As the German pathologist Rudolf Virchow in 1855 stated: "omnis cellula e cellula" ... "Wherever a cell develops, another cell must have existed." Improving microscopic methods and staining techniques allowed the structure of both plant and animal cells to be studied. For example, it was discovered that cells of higher organisms contain a cell nucleus, which itself divides prior to the division of the cell. In addition, the nucleus contains structures called chromosomes. At the beginning of the twentieth century, experimental and cytological examinations postulated that it must be the chromosomes that carry the genes because chromosomes and genes are distributed in the same manner from generation to generation.

1. Cells and Cell Division

The cell is the smallest unit of an organism. Every cell consists of an outer membrane, a nucleus in its interior, and is filled with cytoplasm. The membrane is built by a double layer of water-repellent lipid molecules together with numerous proteins. The cytoplasm is a viscous liquid containing a high concentration of various proteins, fatty substances, carbohydrates, and salts.

The most important cell components are the proteins, which function as enzymes or as building blocks for membranes, supportive structures, and other structural elements. Not every cell contains every protein, as many of them are produced only when needed and in amounts that differ from cell type to cell type. The cytoplasm includes various structures that assume different biochemical functions in the formation and decay of cellular components. Ribosomes consist of various proteins as well as ribonucleic acid (RNA) that are orderly arranged and play a large role in the synthesis of proteins (see Chapter III). Several organelles create spaces enclosed by membranes where enzymes present in high concentrations can fulfill specific biochemical processes. Interspersed throughout the cell are mitochondria that generate energy for the cell (Figure 2).

The nucleus, encased by its own membrane, is filled with a dense structure named chromatin, so called because it can be stained easily. Before and during the division of the nucleus, chromatin condenses into distinctive, usually threadlike structures named chromosomes. Before cell division, the chromosomes are duplicated lengthwise, and then they are distributed one each to the daughter cells (Figure 3). The centrosome, an organelle in the cell, is particularly important during this process. It is the first structure to divide, with each new centrosome migrating to the opposite poles of the cell. With the centrosomes as the starting point, stellate fibers arranged like a spindle connect to the chromosomes. When these spindle fibers contract, they each pull one of the chromosomes to one of the poles. Each daughter nucleus, therefore, receives one copy of the chromosomes in the original cell. The new nuclei are separated and the two daughter cells finally divide between them. This mechanism of cell division is called mitosis and ensures that the daughter cell receives the same chromosomes as the mother cell.

Animals differ with respect to the number and the types and shapes of cells they have. The worm *Caenorhabditis elegans*, for example, has exactly 959 body cells composed of about 10 different types such as muscle cells, nerve cells, and skin cells. Most animals, however, are composed of a variable number of cells. Human cells, for example, come in more than 200 different shapes and functions. It is interesting to note that smaller animals actually have fewer cells, and,

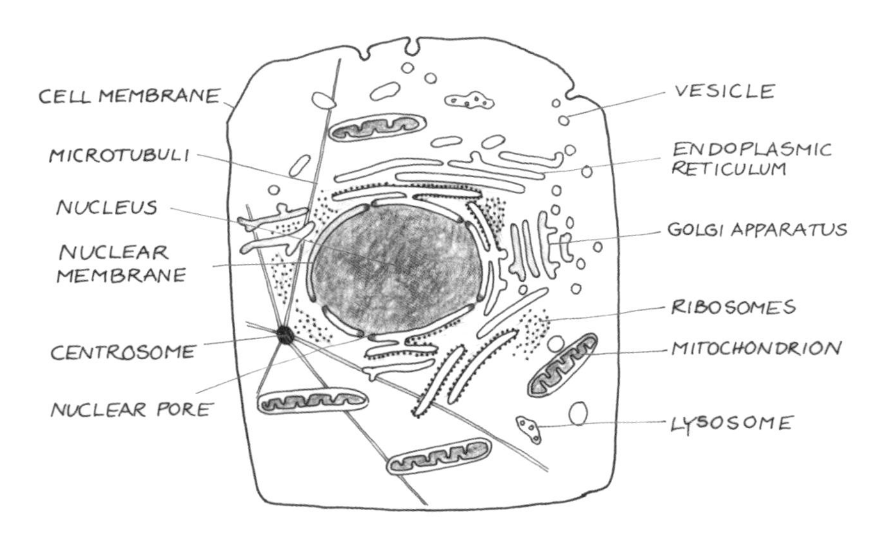

Figure 2. *An Animal Cell.* The cell is surrounded by a membrane, which, via vesicles can exchange material with the outside. A double membrane punctuated by nuclear pores surrounds the nucleus. These pores are highly organized protein complexes which regulate traffic in and out of the nucleus. The endoplasmic reticulum, associated with ribosomes, and the Golgi apparatus are compartments where proteins for excretion are produced. Mitochondria contain membranes with a large surface. In the mitochondria, energy for the cell is generated from nutrients. Microtubules are the building blocks of the cytoskeleton; they originate at the centrosome. Organelles, like the lysosomes, rid the cell of materials that are superfluous or even harmful.

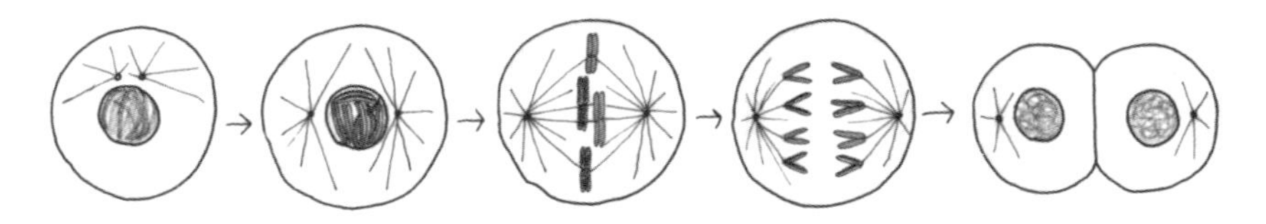

Figure 3. *Cell Division.* For simplicity, only two pairs of chromosomes are shown (red and gray). First the chromosomes are doubled and a spindle forms. Then the nuclear envelope breaks down and the spindle microtubules attach to the chromosomes. Next the chromosomes are pulled into opposing halves of the cell so that they separate into two chromatids per chromosome. Finally, the cell divides and the nuclear envelope reconstitutes itself.

likewise, larger animals do not simply have larger cells but more of them. The mouse has 2×10^9 cells and the human has 3×10^{13} cells. This implies that there is a certain minimum size for cells and that cells cannot be built any smaller. Mammalian cells, for example, have a diameter of about 10 micrometers.

2. *Fertilization*

Animal development begins with the fusion of the egg cell and the sperm, commonly known as fertilization. The resulting zygote receives its cytoplasm exclusively from the egg cell, but the nucleus is composed equally of maternal and paternal chromosomes. As embryonic development begins, the fertilized egg cell divides many times. In the early stages of animal development, known as blastula, the cells look the same. Shortly thereafter, however, the cells start to become visibly different, and the embryo takes shape. This important process is called gastrulation. Cells sort themselves into groups, they ingress or fold into the embryo, shift, continue to divide, and form the primordia of the various tissues and organs. Finally, cells differentiate into organs, which fulfill various functions in the animal (Figure 4).

It became clear that heredity is tied to the process of reproduction. Often, plants, and even certain animals such as polyps or sponges, reproduce asexually by budding. It is easy to see that the buds, or layers, develop into the image of their parents. However, all plants and animals of higher order have two sexes. Both sexes produce reproductive cells that are called germ cells, or

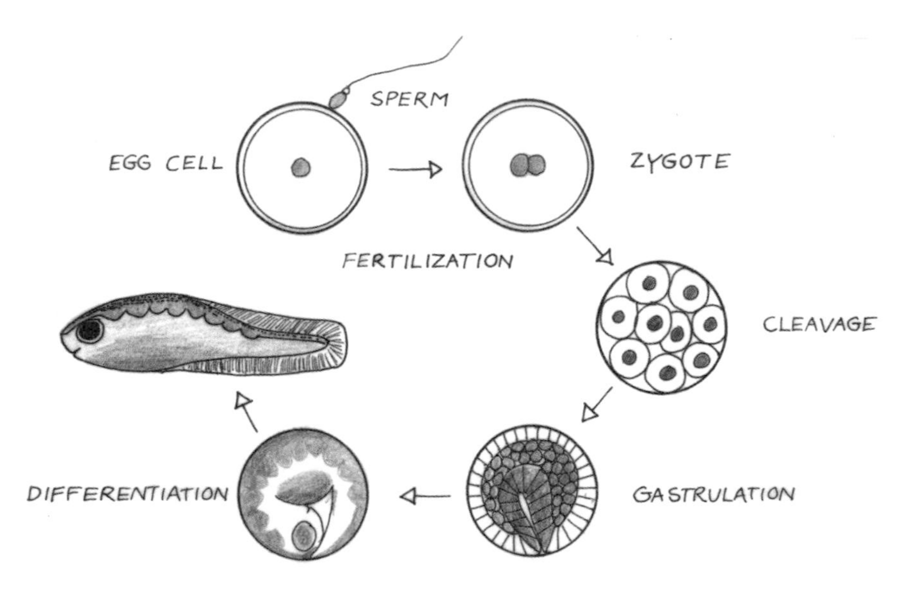

Figure 4. *Development.* A simplified and generalized view of embryonic development, beginning with fertilization. Cleavage follows, dividing the fertilized egg into many small cells. During the process of gastrulation, the overall organization of the body plan is laid down involving cell rearrangements. Then organs start to form and begin to differentiate into various tissues.

gametes. The female egg cells that eventually become the embryo are large and immobile while the smaller male sperm cells are produced in often incredibly large numbers.

The development of a new individual begins with fertilization. The female and male gametes, the egg cells and the sperm cells, each have a distinct shape. The egg cell is rich in cytoplasm and often contains a large amount of yolk that provides nutrients to the embryo up to the moment of hatching. The sperm cell is tiny by comparison, and contains only the nucleus and the centrosome, an organelle not present in the egg cell. Often, the sperm cell also has an organelle for locomotion known as the flagellum, which allows the sperm to swim toward the egg cell.

In marine animals, the fusion of egg cell and sperm cell takes place in the water outside of the body. With the rise of land animals, fertilization adapted to the fluids contained within the body. As soon as a sperm cell penetrates the egg cell, the flagellum of the sperm is shed, and the nucleus and centrosome enter the egg. At the same time, a membrane quickly encloses the egg cell ensuring that not more than one sperm enters. The nucleus of the sperm then approaches the nucleus of the egg and they fuse.

Embryonic development begins with the initial cell divisions. The cytoplasm of the egg cell is distributed to the daughter cells, which become smaller in size as their numbers increase through a process known as cleavage. Later in development, at the point when the organism is no longer confined to the size of the original egg cell, the division of the nucleus is preceded by a growth of the cytoplasm so that each daughter cell may become as big as the mother cell.

Certain aspects of fertilization deal specifically with the role of the chromosomes, the centrosome, and the cytoplasm. Usually, the egg cell can only develop if fused with a sperm cell. One interesting finding was that the nucleus of the sperm cell is not necessary for the early divisions at the very beginning of development. The centrosome, which is also part of the sperm, however, is essential for the initial divisions. But there are some cases in which a centrosome can actually be created within the egg. This process is called parthenogenesis, or virgin fertilization, as development begins without fertilization. Male bees, for example, develop through parthenogenesis.

But for some time one question remained. Within this process of fertilization, where were the carriers of hereditary factors located—in the cytoplasm or in the nucleus? Clever experiments helped find the answer. It had been common knowledge that when crossbreeding two different species, the hybrid would show both maternal and paternal characteristics. An exchange of parents did not result in major differences even though the cytoplasm was

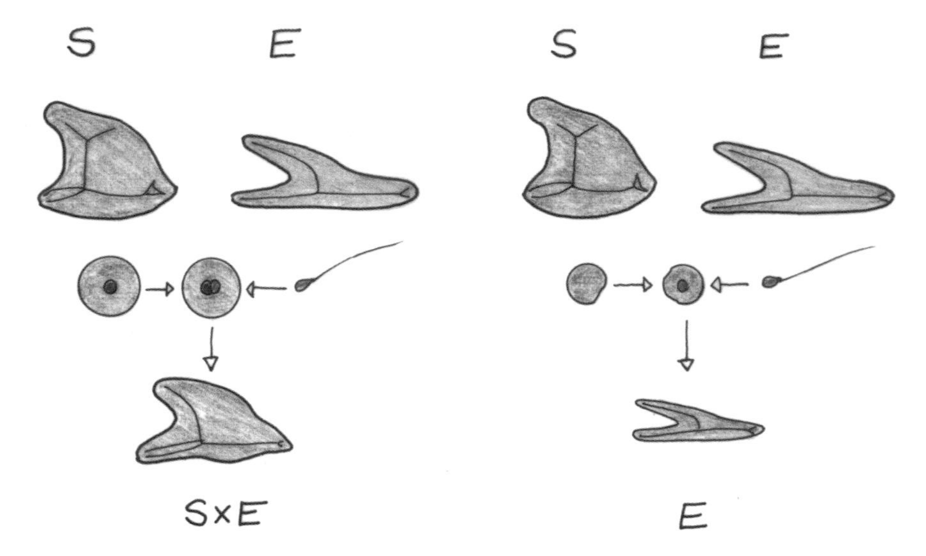

Figure 5. *The Importance of the Nucleus.* When two different species of sea urchins (S and E) are crossed (left panel), the resulting hybrid larva shows a mixture of maternal and paternal qualities. By heavily shaking the eggs, Theodor Boveri could produce eggs without a nucleus. If such eggs are fertilized with the sperm of another species (right panel), larvae develop that only resemble the father even though the cytoplasm was entirely maternal.

exclusively maternal, in other words, solely from the egg. The German zoologist Theodor Boveri (1862–1915) observed that when clumps of cytoplasm without any nucleus were fertilized with the sperm of a different species, the resulting mini larva looked like the father and not like the mother. This proved that it is the nucleus rather than the cytoplasm that carries all genetic information (Figure 5).

3. Chromosomes and Genes

The American biologist Walter Sutton (1877–1916) examined the development of germ cells. He showed that a certain species of grasshopper had 11 chromosomes that were visibly distinct from each other. Furthermore, he showed that within the body cells the chromosomes appear in duplicate, termed the "diploid" state. During the cell division that precedes the development of egg cells and sperm cells, the corresponding homologous chromosomes arrange in pairs and are distributed one each into one of the daughter cells by the dividing spindle. Which copy of the two homologous chromosomes is distributed into which daughter cell is left to chance. Such a division results in daughter cells that contain only one copy of each chromosome, and are thus called haploid.

This form of cell division is called reduction division or meiosis in contrast to the normal division called mitosis. In the course of fertilization, the haploid cell nucleus of the maternal egg cell fuses with that of the paternal sperm cell. The result is named a diploid zygote. Mitotic divisions result in the multiplication of diploid cells, whereas meiotic divisions reduce the chromosome number back to the haploid state during the formation of egg and sperm cells.

The distribution of chromosomes from generation to generation corresponds exactly to the distribution of hereditary factors according to Mendel's laws (Figure 7). Walter Sutton's theory predicted that there should only be as many independent hereditary characteristics as there are pairs of chromosomes. In 1902, he published the possible combinations in a table. The table is relatively easy to read with just two chromosome pairs. For example, there are 2 × 2 totaling 4 different combinations in the haploid nucleus, and 4 × 4 totaling 16 different combinations in the zygote. If there are 10 pairs of chromosomes there are more than a thousand possible combinations in the egg or sperm cell, and 1 million in the zygote. Humans have 23 pairs of chromosomes.

Is it only the number of chromosomes that matters or are chromosomes distinct from each other? If so, what are their functions? Theodor Boveri discovered that at least one copy of every chromosome is necessary for the normal development of the animal. If one or more chromosomes were missing completely, characteristic developmental problems resulted.

The experiment that led him to this conclusion was an examination of sea urchin embryos developing from eggs that had been fertilized by two sperm, something that happens only very rarely (Figure 6). The egg cell then receives four spindles that during the first division form four daughter cells simultaneously. But since every chromosome is present in three copies—one from the egg cell and one each from the two sperm cells—the developing cells would vary in the numbers of chromosomes they receive.

Instead of receiving two copies of each chromosome, Boveri showed that many cells have only one, others three, and others miss that particular chromosome altogether. These descendant cells could not develop normally and showed a characteristic set of defects. This means that chromosomes are individually different from each other and that every one of them carries specific genes for specific functions in the development of the animal. Therefore, genes from different chromosomes cannot substitute for each other.

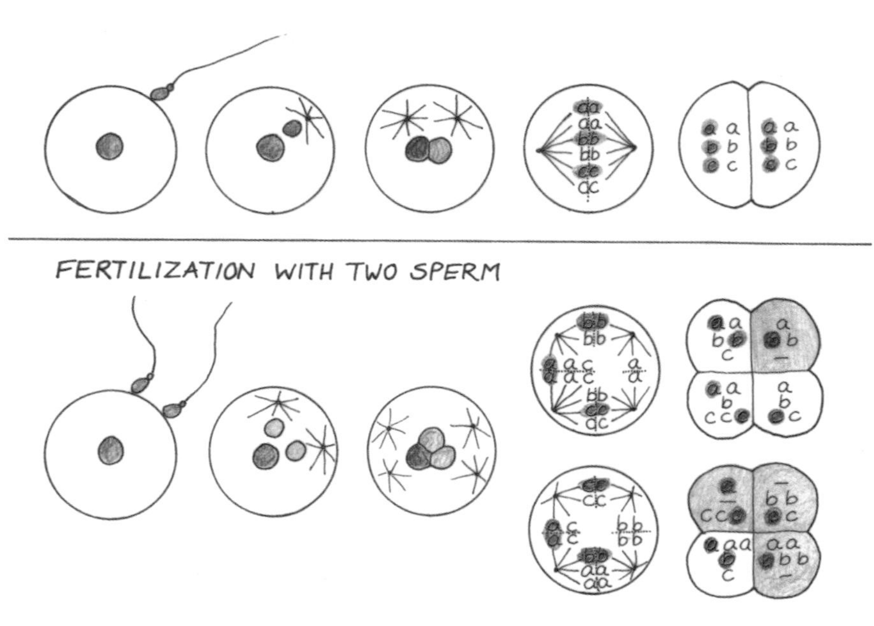

Figure 6. *Double Fertilization.* During normal sea urchin fertilization (upper panel), the spindle that develops between the centrosomes distributes the chromosomes of the maternal (red) and paternal (gray) nucleus to the daughter cells such that each cell receives one paternal and one maternal copy of each chromosome. If two sperm fertilize the cell at the exact same time, four spindles will develop (lower panel). In this case, the three sets of chromosomes (one from the mother and two from the fathers) are distributed among these four spindles. Four cells are created at the same time containing varying numbers and combinations of chromosomes. Many cells do not develop normally. From the frequency of defective cells, Theodor Boveri concluded that these cells (gray) were missing one chromosome altogether and, thus, that the chromosomes must be different.

Sutton and Boveri published their results more than 100 years ago, in 1903. Their results can be summarized as follows: the chromosomes carry the genes that are present in duplicate within the body cells, yet are singular within the gametes. Only one of the pair of homologous chromosomes is passed on to the offspring; which one of the two is left to chance. All body cells contain two copies of all genes, one from the mother and one from the father (Figure 7).

4. *Germ Line and Clones*

The discovery that all cells within an organism contain all chromosomes, and thus all genes, was of crucial importance to the study of how cells become

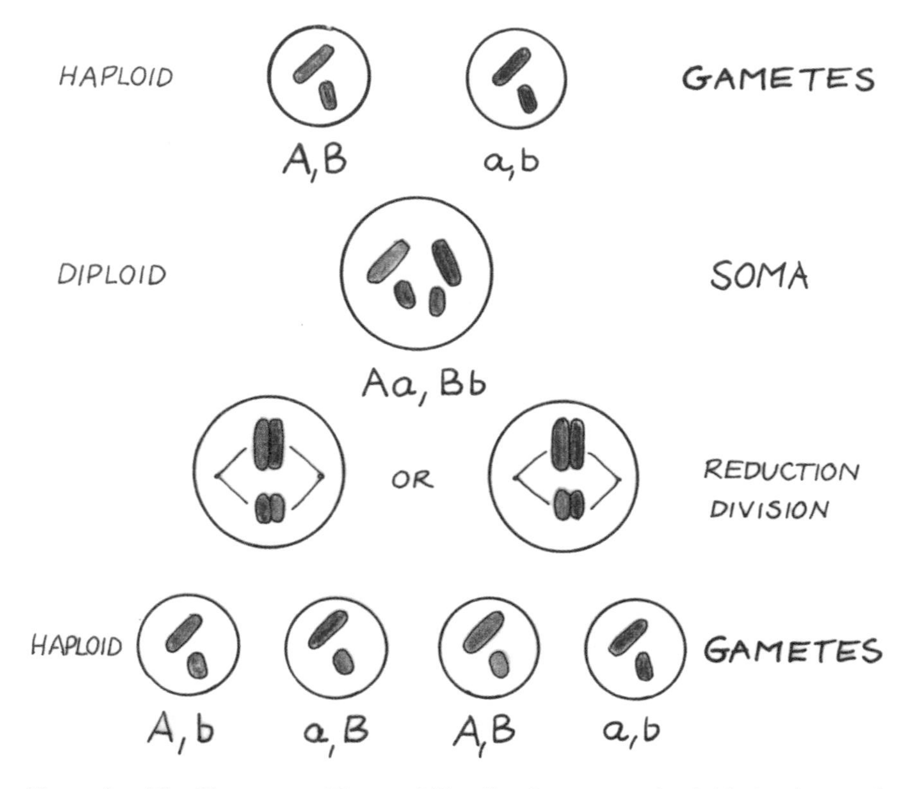

Figure 7. *The Chromosome Theory of Heredity.* Gametes are haploid, they have only one copy of each chromosome, while the diploid body cells, called soma, have two copies, one from the father, one from the mother. For simplicity only two chromosome pairs are illustrated here. When gametes develop, the homologous chromosomes arrange in pairs and are distributed into the gametes independently from each other. This explains Mendel's Laws. This figure does not take into account the phenomenon of recombination (see Chapters III and VI) that was not discovered until later.

different from one another. Previously, the famous yet incorrect thesis of the German embryologist August Weismann (1834–1914) had postulated that the differentiation of cells was caused by an unequal distribution of genes to the daughter cells. In contrast, we now know that an organism is composed of a large number of genetically identical cells. Thus, it is a clone that originates from one founder cell, the zygote.

The observations of Boveri and Sutton may well have sprung from isolated cases, but in the 1960s, the British researcher John Gurdon (1933–) tested and confirmed their theses. He transplanted cell nuclei from differentiated cells using the gut of a tadpole into egg cells from which the nucleus had been removed. In rare cases these composites developed into a normal tadpole (Figure 8). With this experiment John Gurdon proved that the differentiated gut cells still contain all genes required for the development of a tadpole. The resulting tadpole is identical to the tadpole that donated the gut cells—it is indeed its clone. This way of cloning animals by nuclear transplantation has also been shown to work with other animals, such as sheep. However, the very low success rate of this method raises the question: why is cloning so difficult?

In higher animals, specific cells are responsible for the production of gametes. These "primordial germ cells" contain distinctive components that are present only in this cell type. The germ cells separate from the other body cells very early on during development and migrate to the sexual organs, the gonads, where later they form egg or sperm cells. In flies, for example, the first cells that develop are named the pole cells, and they are the ones that will develop into the germ cells (Figure 20). The gametes and their predecessors are called the germ line, as opposed to the body cells, which are called the soma. The germ-line cells go through a special developmental process, including most likely a program protecting the genes from mutations and thus ensuring that the cells responsible for the next generation remain as unharmed as

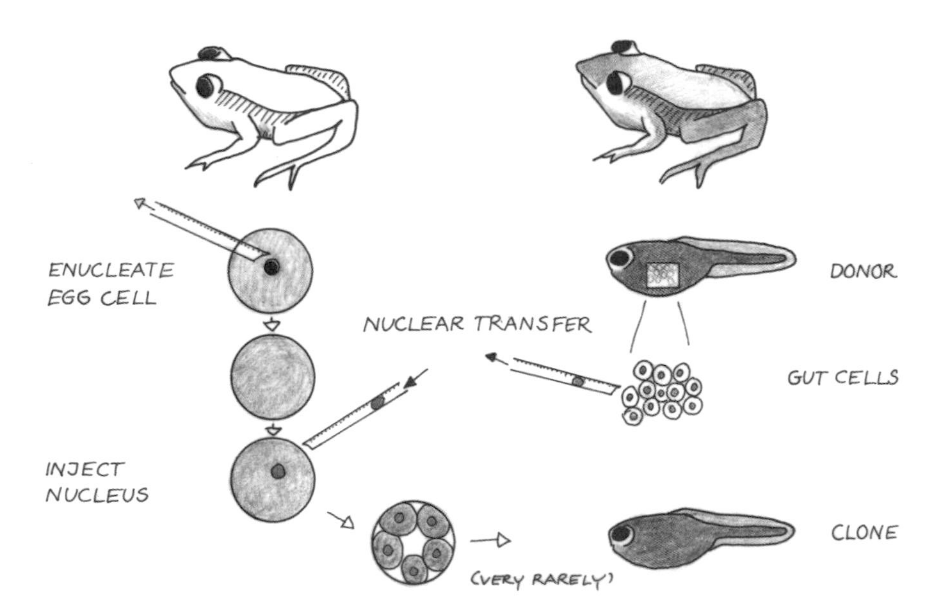

Figure 8. *Cloning.* The nucleus of the egg of an albino frog is removed. It is replaced by the somatic cell of a differentiated tadpole from a normally pigmented frog. The tadpole developing in rare cases from this experiment is normally pigmented and thus has the genotype of the animal from which the nucleus was derived. It is its clone.

possible. It is not yet clear what this program entails exactly, but it is clear that the germ line is necessary as it contains characteristics that the soma does not have. Because any offspring will develop solely from the gametes, changes to somatic cells have no influence on the offspring's hereditary information. In addition, gene mutations will only be inherited if they take place in the germ-line cells. This is why acquired properties are not inherited.

Asexual reproduction is common among plants, but rare among animals. The freshwater polyp *Hydra* reproduces by budding, and some insects, like aphids, show life phases of quick multiplication through diploid eggs that form large, genetically identical clones. But in difficult times, even these animals reproduce sexually. Sexual reproduction leads to more variations among the descendants, of which some may have a greater chance to survive.

5. *The Influences of the Cytoplasm and the Environment*

If all cells have all genes, the origin of differences arising during development of an organism must reside in the cytoplasm. Factors in the cytoplasm decide the fate of the cells by determining which genes are active and which are not. The two daughter cells that originate from the first cleavage division of a frog can produce a viable, albeit smaller, frog. But just a few divisions later, this is no longer possible as the individual cells do not have enough cytoplasm and the cytoplasm that they do have differs between the individual cells (Figure 9). Theodor Boveri observed that sea urchin eggs have an intrinsic polarity and that an artificial splitting of the eggs at right angles does not produce complete embryos. With regard to their ability to produce an entire embryo, the right and left sides of the egg are identical, but the top and bottom are different. He concluded that cytoplasmic factors are at play and that they gradually increase or decrease in concentration from top to bottom of the egg in determining the developmental fate of the cells. In other words, there are differences within the cytoplasm that determine the fate of the daughter cells.

The secret of embryonic development is the control of gene activity in time and space. Together with the factors in the cytoplasm, the genes constitute a building plan for the developing organism, a blueprint that is realized step by step. As explained earlier, the cytoplasm originates from the egg. But the cytoplasm also receives signals and information from the environment, including the neighboring cells. This information is transmitted to the genes in the nucleus. In this manner, the fate of a cell is dependent on both the cytoplasm and external influences.

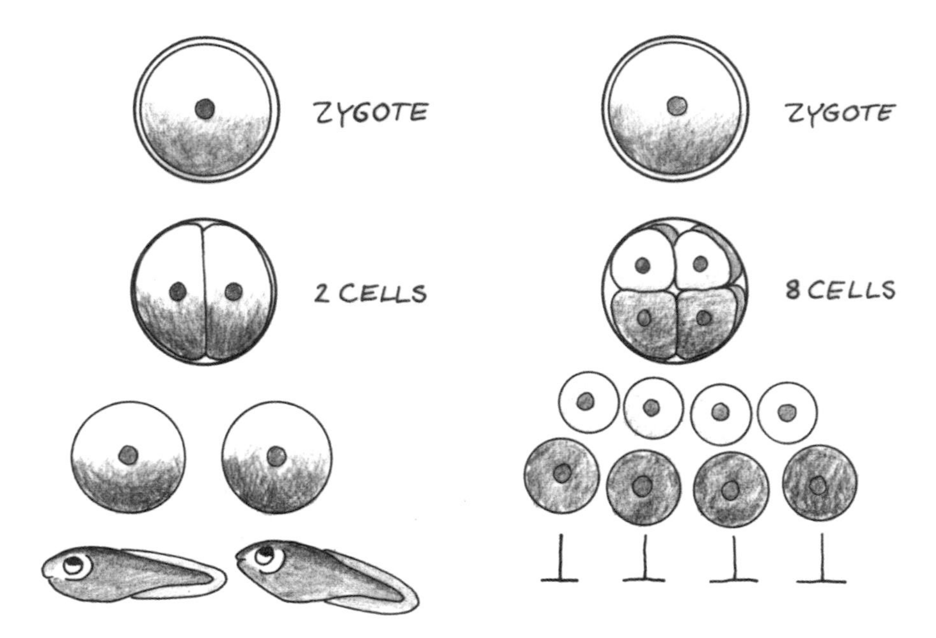

Figure 9. *Multiples.* The separation of cleavage cells at the two-cell stage can produce two normal offspring. With sea urchins, this is even possible at the four-cell stage. At later stages (right panel), this is no longer possible because the cells do not contain all necessary cytoplasmic factors any longer and are too small.

Researchers like Theodor Boveri clearly realized that in their time it would not be possible to find out what genes are, nor what factors influence development. It took nearly a century before modern molecular genetics was capable of identifying and isolating these factors.

From an historic standpoint, it is interesting that at the beginning of the twentieth century, embryology and genetic research began to develop as separate disciplines. One important reason may have been that the organisms most suited to one field of study were not particularly useful to the other. Amphibians, such as frogs and newts were excellent for embryological research, as they lay huge eggs 2–4 millimeters in diameter that could easily be surgically manipulated. For example, experiments in which specific regions of the embryo were isolated and recombined led to the discovery of a special region in the newt egg that came to be called the organizer.

The organizer has a long-range influence on its neighboring cells and can even induce the formation of an additional body axis. In the famous 1923 experiment by the German zoologist Hans Spemann (1869–1941), the organizer cells from one embryo were transplanted into another embryo of the same age, in the region that develops normally into the stomach. But after the transplantation, a second head and trunk region developed instead of stomach. Since the two embryos were from two different species and easily distinguishable by way of their pigmentation, it was possible to determine that the new axis did not originate from the donor tissue but rather that the donor tissue had influenced the cells in its new environment to form an additional body axis. This crucial experiment prompted attempts to isolate and biochemically characterize the factor operating as the organizer. But despite great efforts, it took 70 years before the gene responsible was isolated and identified.

Aside from some frustrations and false interpretations along the way, several quintessential discoveries had been made at the dawn of the twentieth century. One was the realization that genes are the discrete units of heredity that guide development. Another was that the cytoplasm exerts influence by determining which genes are active and which are not. And a third was that the nucleus contains, with the genes, the entire hereditable information of a developmental program. With these discoveries, the nature of genes and their regulation became the main concern of biology.

CHAPTER III

Genes and Proteins

GREGOR MENDEL DID NOT YET USE THE TERM GENE. HE REFERRED INSTEAD to features and elements, terms that come close to our modern understanding of phenotypes and genes. Genes are the units of heredity and are noticeable only if they appear in different states or alleles that change a particular feature in the appearance of the organism. He intentionally restricted his experiments to simple cases with only two visibly distinguishable alleles, the wild type and the mutant. The hereditary patterns that emerged were easily recognizable and led him to conclude that every gene appears in duplicate, with one copy coming from the mother and one from the father. Furthermore, he proposed that they are passed on to each subsequent generation complete and undivided. The validity of his laws was confirmed more than once during the year 1900. Subsequent research shortly thereafter revealed the connection between Gregor Mendel's elements and the chromosomes.

The material composition of genes was still a mystery in the first part of the twentieth century when the terms genetics, gene, and genotype were coined. Genes could not be seen as such, but conclusions could be drawn regarding their mode of action through the process of breeding individuals with various distinctive features. Indeed, intense studies on the fly *Drosophila melanogaster* gave insight into how genes are passed from generation to generation, as well as to the organization of these genes on the chromosomes. At first, these studies concentrated on mutants with a clearly visible adult phenotype, such as changes in eye color. They confirmed that Gregor Mendel's laws were also applicable to animals and, hence, were universal. In addition, this work also discovered the sex chromosomes and revealed the processes of genetic linkage and recombination. Before long it was possible to map the relative positions of genes along chromosomes.

Yet the question of what a gene is exactly and what it does was not easy to answer. The mutant phenotypes offer vital information about the function of a gene. For example, a mutation in the white-gene results in flies with eyes that are white rather than the usual red, demonstrating that the white-gene has a role in generating the red eye pigment, a function that has been lost in the mutant. However, there is a whole host of other genes such as pink, claret, scarlet, cinnabar, and brown—all of which when mutated affect the fly's eye color. Therefore, it is evident that any given quality is under the influence of more than one single gene. The converse is also true as most mutations result in complex phenotypes where more than one feature is affected. Although research on *Drosophila* was successful in revealing many important hereditary principles, the biochemical nature of genes and the way they work were unraveled in studies of much simpler single-celled organisms such as bacteria and fungi.

1. The Genetics of Drosophila

The American biologist Thomas Hunt Morgan (1866–1945) demonstrated that the common fruit fly *Drosophila* was an ideal subject for the study of genetics. As is true for all insects, the fly has an outside skeleton or a cuticle that is covered with hairs and bristles that form in a consistent pattern. Most importantly, it has a large number of prominent features: red eyes, brown body, characteristic antennas, and wings, structures that are easily visible under a magnifying glass so that even the slightest changes caused by mutations can be recognized. Furthermore, the generation time is short and one breeding pair can produce many offspring.

The fly larva, commonly known as the maggot, is a segmented, worm-like creature that develops inside the egg with a shape much simpler than that of the adult fly. The larva grows and sheds its cuticle twice before forming into a pupa. Then the mature fly, or imago, emerges from this pupa after a 12-day period of development (Figure 10).

Sex Chromosomes. In 1910, Morgan described a mutant strain of flies with white eyes. Experiments with these mutants brought about extraordinary findings. Since the dominant allele causes the eyes to be red, the normal eye color of the fly, crossbreeding homozygous white-eyed females with red-eyed males should have resulted in exclusively red-eyed offspring. However, only female offspring developed red eyes whereas all of the male offspring had white eyes (Figure 11).

As it turns out, the chromosome with the white-gene is present twice in females and only once in males. This was identified as the X-chromosome. Instead of a second X-chromosome, males have a Y-chromosome with substantially fewer genes. Even though the Y-chromosome is not homologous

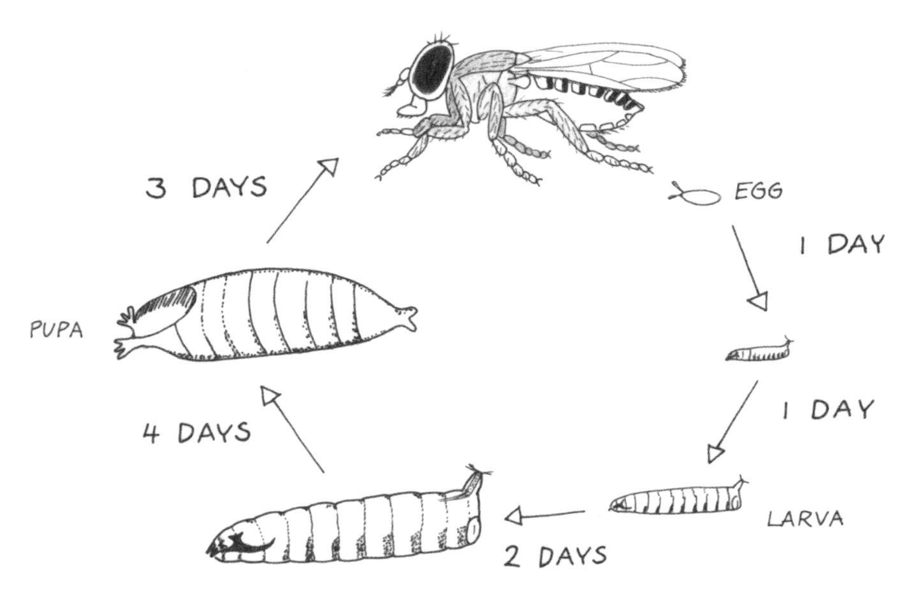

Figure 10. *The Life Cycle of Drosophila.* This well known fruit fly is only a few millimeters long. The fly lays its numerous eggs in ripe fruit on which the simple larvae feed. Fly larvae have neither heads nor limbs. The larva grows fast, molts twice, and then migrates to a dry place where it enters the pupa stage. After 11 days in total, an adult fly complete with legs, red complex eyes, antennas, and wings, emerges and flies away.

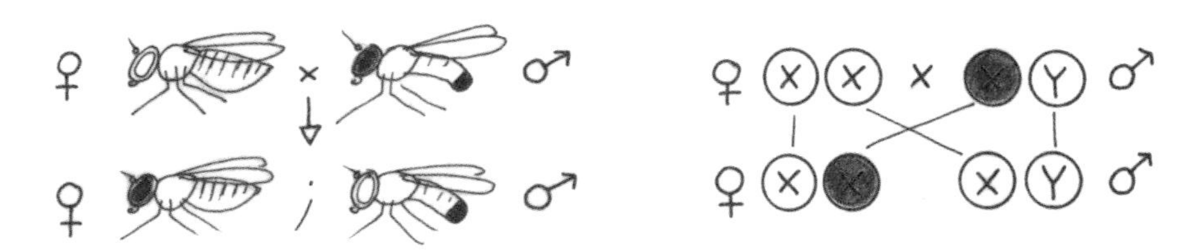

Figure 11. *X-Chromosome.* When crossing mutant white-eyed female flies with normal red-eyed male flies (left panel), red-eyed females and white-eyed males will result. This can be explained by the fact that the mutation of the white-geneis located on the X-chromosome that in males occurs only once. Males receive their X-chromosome exclusively from their mother and the Y-chromosome from their father, as shown on the right.

with the X-chromosome, they still pair together during meiosis such that the sperm receive either an X- or Y-chromosome (Figure 11). Males receive their X-chromosome exclusively from their mothers. Furthermore, as they contain only one X-chromosome, the phenotype of a recessive mutation in a gene that is located on the X-chromosome can easily be seen in the male fly.

Also, in humans, the chromosome combination of XX and XY defines females and males, respectively; however, this is not generally the case. For example, female birds have two different sex chromosomes, whereas male birds have only sex chromosomes of one kind. In addition, the sex of some animals, like the crocodile, is determined not by chromosomes but by temperature. When chromosomes determine sex, the offspring will normally contain as many males as females. This is not the case, however, if sex is determined by temperature, as at extreme temperatures all individuals will be of the same sex.

Recombination. Before long, more mutations changing various characteristics of the fly's body were found. These included the length of the wings, the color of the body, the number and position of bristles, and the structure of the eyes. Crossbreeding showed that most of these genes reside on one of the three large chromosomes, which led to the discovery of yet another deviation from Gregor Mendel's laws. If two genes are on the same chromosome, their qualities are inherited together: they are linked rather than independent. This linkage had been previously overlooked in other organisms. This is because the probability of two genes lying on the same chromosome and therefore being linked is rather high in the case of the fly, as it has only four chromosomes. The probability is rather low with most other animals, as they have a much larger number of chromosomes. For example, the sea urchin studied by Theodor Boveri has 18, the zebrafish has 25, and humans have 23 pairs of chromosomes.

However, even this new exception to Mendel's laws had its own exception. Two mutations on the same chromosome are not always linked together. The reason for this is due to the process known as genetic recombination. During meiosis, when homologous chromosomes pair, entire chromosome sections may be exchanged in a process called crossing over. Crossing over results in the haploid egg, or sperm cells, receiving one chromosome that is a mixture of the two chromosomes of the initial pair. This increases the number of possible combinations of alleles within the gametes beyond those already possible just from the combinations of the different chromosomes themselves. While these crossing over events essentially take place at random positions along the chromosome, the closer two genes are on the chromosome, the more rarely a crossing over will occur between them. This means then that the frequency of crossing over between two genes gives some idea of their proximity on the chromosome (Figure 12). These findings made it possible to delineate genetic maps representing the order of genes on the chromosome.

Giant Chromosomes. A fly larva grows not by increasing cell number, as in many other animals, but by increasing cell size. The chromosomes double as

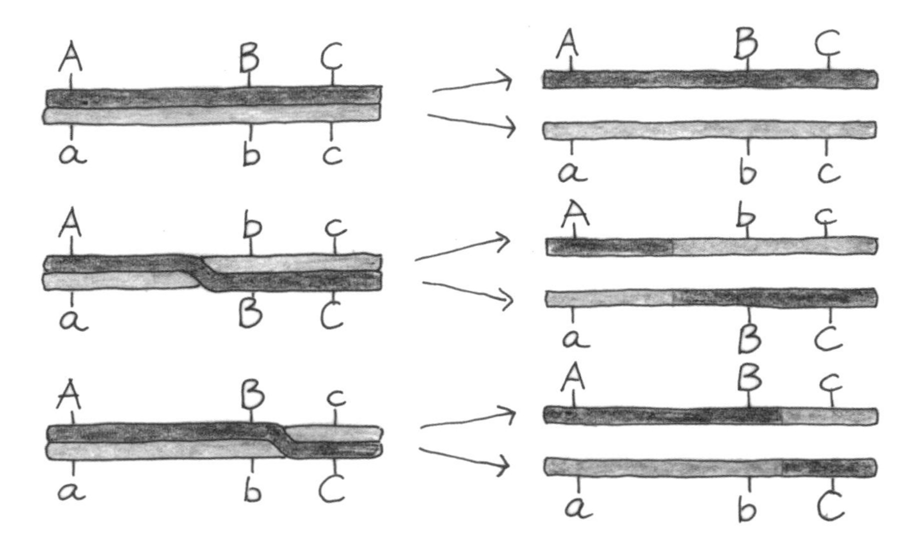

Figure 12. *Recombination.* At the beginning of meiosis, the chromosomes align and crossing over takes place. Sometimes, parts of one chromosome are rejoined to the other. This process results in a recombined chromosome that is partly derived from the mother and partly from the father (lower two panels). In these cases, alleles on the originally maternal and paternal chromosomes are newly combined. Recombination is more frequent between genes that lie further apart on the chromosome, in this case it is more frequent between B and C than between A and B. Therefore, the frequency of recombination between two genes is a measure of their relative distance on the chromosome (see also Figure 39).

usual and the cytoplasm grows in volume accordingly, but the cells themselves do not divide. Instead, the cells become gigantic and polyploid, containing thick bundles of multiplied chromosomes, which take on a characteristic appearance of banded patterns. These bands are particularly well developed in the giant chromosomes of the salivary glands. Even though the bands do not exactly correspond to the individual genes, their patterns can correlate the genetic maps to actual structures on the chromosome. Thus, these bands allowed genes to be visible for the first time (Figure 13).

2. *Mutation*

The hereditary change of a gene is called mutation and animals carrying such genes are called mutants. The condition of the alleles of one or more genes is called genotype and the appearance in the organism is its phenotype.

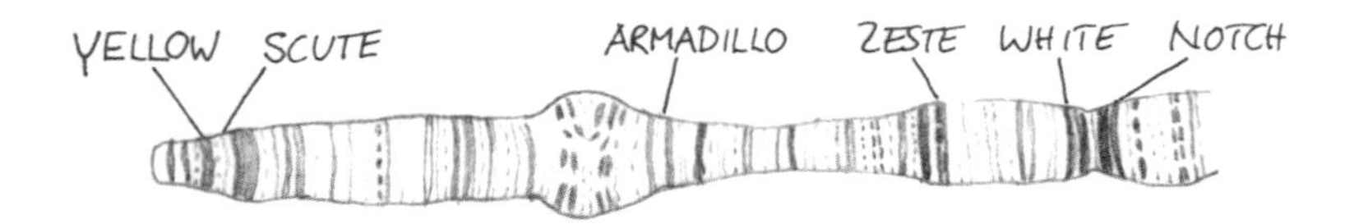

Figure 13. *Giant Chromosomes.* In some tissues of *Drosophila*, the chromosomes become so big that they display a banded pattern. Shown here are bands of the tip of the *Drosophila* X-chromosome that were described and numbered by American geneticist Calvin Bridges (1889–1938). This figure covers about 150 of the 5,000 bands of the *Drosophila* genome and indicates the location of some important genes.

Spontaneous mutations are quite rare and how often they occur depends on several factors. For example, x-rays may result in breaks of chromosomes that only heal partially or in entire chromosome segments being lost. The extent of such a deletion can be seen in the giant chromosomes. Most mutations are the result of the gene having been damaged in such a way that it can no longer fulfill its normal function. Such null mutations behave as if the gene had been lost altogether. With the aid of certain chemicals, scientists can increase the frequency of spontaneous mutations.

The symbol (+) is used for the wild or normal type of the gene, and the symbol (–) is used for the mutation. The person discovering the first mutant of a gene usually names it by using a simple description of the phenotype. A gene receives its name from the mutant rather than from its normal form. For example, instead of the normal red eyes, a mutation in the white-gene causes flies to have white eyes and a mutation in the pink-gene causes flies to have pink eyes. The allele of the wild type is dominant in most cases, whereas null mutations are mostly recessive. Therefore, a single copy of the intact gene is sufficient to fulfill its normal function, that is, a fly with the genotype $w+$, $w-$ will have red eyes. There are also weak mutations that still carry out some of the normal functions of the gene. In very rare cases, a mutation does not abolish the gene's function but rather changes its site of activity. If, for example, one gene is placed under the control of another gene, it becomes active at the wrong place. One such mutation, Antennapedia, leads to flies that have a perfectly formed pair of legs sprouting out of their head.

In most genes, a mutation decreases the animal's chances of survival. For example, homozygous mutants in about one-third of the fly's genes do not survive and development arrests during the embryonic, larval, or pupal stages. There are only a few genes that allow the animals to survive and show a visible phenotype such as the white-gene or the yellow-gene that pales the fly's body.

Mutations in more than half of all genes are silent, that is mutant animals survive and do not display a visible phenotype. This is not only true for flies but also for other animals in which the function of a large number of genes has been studied, such as mice. Yet, it is likely that these genes still have an important function during the animal's life, a function that may be less obvious in the protected environment of the laboratory. These very gene mutations may actually be responsible for the significantly diminished fitness in the offspring of animals resulting from inbreeding. Therefore, most animals and plants have behavioral, morphological, or physiological mechanisms that prevent inbreeding. Mutants with a visible phenotype are also often less robust than their wild-type counterparts. Mutations causing drastic phenotypes play virtually no role in the process of evolution because in the wild the striking visibility of mutant individuals severely debilitates them.

At the time of these discoveries it was not possible to reveal the molecular composition and biochemical effect of genes from investigations of *Drosophila*; however, the study of the genetics of much simpler organisms such as fungi and bacteria did finally uncover the biochemical connection between gene and phenotype.

3. The Molecular Nature of Genes

Fungi such as mold and bacteria are haploid, meaning that they have one copy of each gene. Therefore, the phenotype of every gene is displayed directly in the mutant cell and thus allows for a convenient study of the gene's biochemical functions. It was known that bacterial or fungal cells can be easily bred in media and that some mutants would only grow if the medium was enriched with nutrients such as glucose or with precursors for the synthesis of glucose. This means that such mutants lack certain enzymes necessary for individual steps in metabolism, and that the respective gene correlates with an enzyme. The theory at the time was that every gene produces an enzyme, or in more general terms, that proteins are produced by genes.

The chemical nature of genes remained unclear until the mid-twentieth century. For a long time, the debate was centered on the question of whether genes were made of proteins or nucleic acid. This debate was settled in 1944 when bacteriologist Oswald Avery (1877–1955) and his colleagues at Rockefeller University proved that genetic material consists of desoxyribonucleic acid, or DNA. It came as a surprise to learn this, as DNA had been known since 1869 when it was discovered by the Swiss scientist Friedrich Miescher (1844–1895) in Tübingen, Germany.

Avery worked with pneumococci, bacteria causing pneumonia, which have a gene with two alleles forming either rough or smooth colonies when they are grown on a plate. These genes could be transferred between colonies in cell-free extracts. Though exposing the extracts to enzymes that attack proteins did not deactivate the genes, they were destroyed immediately if the enzyme DNase was added. A further experiment in 1951 conducted by American biologists Alfred Hershey and Martha Chase showed that when the bacteria were infected with a virus, it was only the DNA of that virus and not its protein that entered the cell.

The DNA Double Helix. DNA is a chain molecule composed of only four different building blocks, the bases adenine, thymine, guanine, and cytosine, all connected by a sugar- (desoxyribose, hence the name) phosphate chain. The interesting aspect of DNA is not its chemical composition, but rather its three-dimensional molecular structure. This structure, the famous DNA double helix, was elucidated in 1953 by the biologist James Watson (1928–) and the physicist Francis Crick (1916–2004) at the Cavendish Laboratory in Cambridge, England. Their model was based on clever combinations of various observations on the chemical nature and physical dimension of the bases and the sugar–phosphate backbone, which they tried to fit into a three-dimensional structure. This structure was then confirmed by the measurements of the physical parameters of DNA by British chemist Rosalind Franklin (1920–1958) at Kings College, London. The DNA chain consists of two strands that run in opposite directions and are arranged in a way that looks like a twisted ladder. The bases in the two strands are facing each other at the inside in a way that the order of the bases on one strand determines the order on the other. For example, adenine in one strand is always opposite thymine in the other strand, and guanine is always opposite cytosine. This order is determined by the special chemical affinity of the bases, which can form stable pairs only as described. The Austrian chemist Erwin Chargaff (1905–2002) at Columbia University had previously discovered that the DNA of any organism contained as much adenine as thymine, and as much guanine as cytosine. This was elegantly explained by the complementary double-strand structure of DNA (Figure 14).

Considering the fact that they consist of only four building blocks, all genes have practically the same chemical composition. Gene function therefore must be determined only by the exact succession of the four bases. It is precisely this seeming simplicity that makes the genetic structure so wonderful. It is a language of four letters that can be read faultlessly in order to make working copies and eventually to replicate precisely.

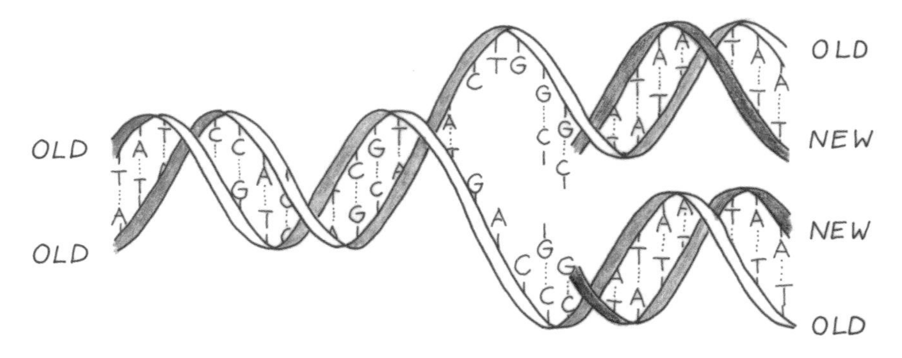

Figure 14. *DNA Double Helix.* DNA molecules carry the genetic information of all organisms. Two strands running in opposite directions form a double helix. The sequence of bases (A, C, G, T) on the two opposing strands is complementary and the bases bind to each other, such that A will always form pairs with T and G with C. The composition of the two complementary strands implies a simple copying mechanism whereby each of the "old" strands serves as a template for the addition of a new one. The nature of the base pairing ensures that the genetic information is faithfully copied. This figure has been redrawn from James D. Watson's book, *Molecular Biology of the Gene*. San Francisco: Pearson Benjamin Cummings, 1965.

4. Protein Synthesis and the Genetic Code

The sequential order of the bases within the genes determines the sequential order of amino acids within the proteins (Figure 15). This translation rule was discovered not long after the elucidation of the DNA structure itself.

The complementary nature of the two DNA strands makes it possible to copy a single strand of individual genes in a process called transcription. It is important to note that only one of the two strands, the so-called sense strand, is copied. The copy is no longer DNA, but rather RNA, a nucleic acid in which the sugar desoxyribose is replaced with ribose and in which the base thymine is replaced with uracil. This single-strand RNA copy of the DNA, called mRNA (m = messenger), serves as a go-between for protein synthesis. The protein is formed in the cytoplasm along the RNA by connecting amino acids according to the given sequence of the bases in a process called translation.

The individual components of this complex synthesis are sorted on the ribosomes and include starter proteins, elongation enzymes, mRNA, and the adaptor RNA molecules known as tRNA (t = transfer). The latter are specifically folded so that on the one end they expose the base sequence complementary to the mRNA, while on the other end they hold the respective amino acid. In this way the amino acids are arranged in the order determined by the

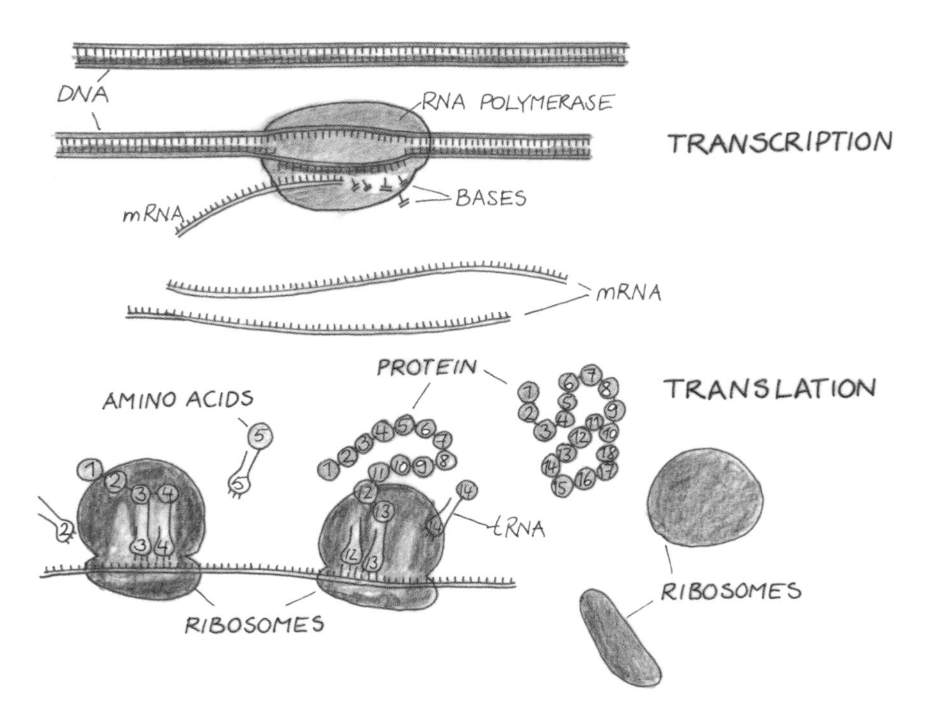

Figure 15. *Transcription and Translation.* During transcription, the DNA strand is unwound and an enzyme, RNA polymerase, builds in compatible bases for the synthesis of single strands of mRNA. As shown below, in the process of translation these mRNA molecules serve as the template for the synthesis of proteins. On the rather large ribosomes comprising two subunits, the tRNA molecules are aligned, together with the mRNA. The tRNA molecules are loaded with specific amino acids depending on their individual codons indicated by the numbers. This genetic code ensures that every three-base codon on the mRNA specifies which type of amino acid is added to the amino acid chain of the protein being synthesized. Enzymes link the amino acids to build a protein chain. In the end, the ribosomes separate and can be used again.

sequence of the mRNA (Figure 15). The amino acids are connected together while the ribosome travels along the mRNA. The growing amino acid chain folds in a characteristic manner. In most cases, the three-dimensional structure of the protein is preordained by the sequence of its amino acids.

The Genetic Code. The sequence of the bases in the RNA copy of the DNA provides the template for the composition of the proteins. Proteins can contain 20 different elements, the amino acids. Each amino acid has its own unique chemical properties. Three RNA bases, referred to as a triplet, determine one amino acid in the protein. With the four bases adenine, thymine, guanine, and cytosine, there are four different possibilities for each of the three positions in a triplet; altogether there are 64 (4 × 4 × 4). But not every single triplet of the 64 different possibilities represents a different amino acid. For example, triplets that are only different in the third base are often translated into the same amino acid. In addition to triplets encoding amino acids, there are triplets that result in the termination of a protein, so-called stop codons. This is the genetic code by which the simple structure of DNA is translated into a virtually endless chemical variety of proteins. Indeed, the proteins, not the genes, are the actual building blocks of the cell. Their biochemical qualities are determined by the succession and composition of the amino acids as well as their three-dimensional structure. In other words, proteins ultimately determine the qualities that create life.

Transcription Control. A gene is called active if its corresponding protein is produced within a cell. The gene's activity can be controlled by blocking or stimulating the transcription of the gene. In many cases, synthesis of the mRNA automatically leads to synthesis of the protein.

In 1962, the French biologists François Jacob (1920–) and Jacques Monod (1910–1976) found DNA control regions in bacteria. Every gene contains a control region in the DNA, known as the promoter, which binds an RNA polymerase enzyme to start the transcription. Near this starter region, other proteins known as transcription factors can either block (repress) or stimulate (activate) the transcription. One example is a protein called repressor, which holds back the RNA polymerase right at the start of the promoter. In bacteria, the control regions are rather succinct compared to the size of the gene, whereas in multicellular organisms the control regions are composed of several elements. These include many binding sites for a number of transcription factors that regulate the complex control of the transcription process. Thus, control of gene activity by transcription factors is a central element in shaping life in time and space.

5. DNA Replication

The complementary nature of the two DNA strands immediately suggested a mechanism for their replication during each cell division. Simply put, unzipping of the strands exposes their bases on each single strand, which then acts as a template for the synthesis of a new complementary strand. The result is two double-stranded molecules that are identical in sequence. In this process, one of the strands remains intact in its entirety while the other one is completely new (Figure 14). This semi-conservative replication was proven in an elegant experiment by Americans Matthew Meselson (1930–) and Franklin Stahl (1929–) in 1958.

In reality, the mechanisms of replication are not nearly as simple as they may seem on paper. First, the strands have to untwist as if untangling a ball of yarn. Then, the DNA molecule has to be cut and later correctly sealed again. Even though there are several repairing enzymes continually checking the DNA for any mistakes that may have occurred during this process, errors can still be introduced. Mutations are one result that may arise from some of these errors if they are not repaired. It is interesting to conclude then that the variations Charles Darwin recognized as the basis for evolution are largely caused by random mistakes during the process of DNA replication. Such errors may change an amino acid in a protein, or if a stop codon is mistakenly created, to a premature termination of the synthesis of the protein. Furthermore, if a single DNA base is deleted or inserted, there will be a frame shift during protein synthesis such that the sequence will be read incorrectly from the point of the error and onward. Sometimes, longer DNA pieces may be incorporated into genes, or duplicated, or lost altogether.

6. Gene Technology

Once the DNA structure had been discovered, the genetic code and the mechanisms for protein synthesis and DNA replication were quickly revealed through studies of the simplest organisms like bacteria and their viruses. In his famous book, *Molecular Biology of the Gene* (1965), James Watson expresses his great satisfaction at having grasped the most important principles of life. At that time, he only wrote of life in its simplest form and not of higher organisms composed of many different cell types, organisms that develop and grow. Nevertheless, it was clear that the genetic code and the foundation of molecular biology he helped to discover, namely that DNA makes RNA that makes protein, also holds true for multicellular organisms. As Jacques Monod said, "What is true for *E. coli* is true for the elephant."

Biochemical experiments shed much light on molecular genetics. For example, they made it possible to isolate and characterize enzymes that read, cut, repair, and then transcribe DNA. Many processes could also be replicated in a test tube, either in a cell-free extract or in a defined enzymatic reaction. Even with these advances, however, the enormous length of DNA molecules made deciphering the function of individual genes a daunting task. The simplest DNA molecules, the circular plasmids which contain only a few genes that multiply in bacteria, still consist of thousands of base pairs.

Unlike proteins, genes cannot be separated according to their chemical qualities, because genes are with regard to their chemical composition virtually identical. And secondly, they are chained in a continuum to other genes as well as to non-gene regions. In other words, the DNA of higher organisms consists of incredibly long molecules that cannot be isolated without breaking them.

Hybridization. Nucleic acids can be compared by means of similarities in their sequence without actually knowing their sequence. These similarities can be detected through a technique called hybridization. It was found that the DNA double strands separate or melt when heated. When cooling down the strands then reunite correctly. Single-stranded DNA or RNA molecules, if complementary, may form mixed, hybrid double-strand molecules. Thus, when mixing DNA and RNA from different origins, strands pair according to their sequence similarities into complementary hybrid double strands. In this way it is possible to see similarities shared between nucleic acids of different origins.

Recombinant DNA. In the 1970s, enzymes were discovered that cut DNA into cleanly defined segments and others that recombined them back together. In this way DNA pieces can also be combined with cut DNA segments from another organism. Such foreign DNA segments can be built into the DNA of plasmids. If such plasmids harboring different DNA segments are introduced into bacteria, the plasmids with these inserted DNA segments are multiplied within the bacteria. A colony originating from a single bacterium containing such a plasmid is called a clone; therefore the process of gene multiplication in bacteria is called cloning. In order to isolate a particular gene or gene segment from the DNA of an organism, collections of such clones are established. The isolation of a particular gene or gene DNA segment is only possible if the collection of bacterial clones is so large that statistically every gene will be present in at least one, preferably several clones. Finding the desired gene in thousands of such colonies is a problem that has to be solved for each individual case. For example, hybridization methods make it possible to isolate clones that carry parts of certain genes out of a gene library. Sometimes these fragments have overlapping segments and they can lead to the isolation of further segments of the gene. In this way, large connected DNA regions that encompass several genes can be assembled.

It is easily possible to produce and isolate many copies of a cloned gene fragment after it has multiplied within the bacteria. In addition, without the aid of bacteria, but rather with enzymes, gene segments can be multiplied in vitro by the method of polymerase chain reaction (PCR).

The Sequence and Structure of Proteins. The sequence of bases of a gene can be determined in pure populations of DNA segments. The genetic code allows the prediction to be made about the composition and sequence of the amino acids of its protein product without having to do the more complicated direct protein analysis. This ability to determine protein sequence by isolating genes advanced biological research by a big leap. In this way the composition of many important proteins was now accessible for study, including proteins present in miniscule quantities and proteins that are unstable and would therefore be extremely difficult to isolate.

Once individual genes have been isolated, large amounts of the corresponding proteins can be produced in bacteria or cell cultures. Rare proteins, such as hormones, blood factors, enzymes, and antibodies, can also be produced in bulk quantities and in particularly pure forms without the need to extract them from human or animal tissue.

The analysis of the sequences of many proteins revealed that there are certain areas or domains that also occur in other proteins with similar functions. For example, several transcription factors contain a domain of 60 amino acids that specifically binds to DNA. This protein domain, called homeodomain, was first discovered in *Drosophila*'s homeotic genes, but it also occurs in other proteins that function as transcription factors. Other transcription factors have different DNA-binding domains. Some domains are present repeatedly within a protein. One of these is the immunoglobulin domain of antibodies that consists of 110 amino acids. Such sequence motifs may also be reliable indicators of where particular proteins will end up once synthesized. For example, proteins that are secreted from cells carry a characteristic sequence of amino acids in the start region of the protein, while other proteins contain sequences anchoring them within the cell membrane. Computerized search programs also help recognize recurrent domains and assist in further understanding their roles.

7. *Genes of Multicellular Organisms*

Even though the way genes work—the complem1entary nature of the DNA strands, the genetic code, the transcription process, and the translation on ribosomes—are basically true for all organisms, the genes of higher organisms show a few unique characteristics. As stated earlier, the majority of the DNA of bacteria and viruses codes for translation into proteins. This is not the case with eukaryotic organisms such as animals and plants. Their DNA is not freely distributed in the cytoplasm, but rather is contained in a nucleus bounded by a nuclear membrane, where it is tightly associated with particular proteins known as histones. These proteins wrap up the DNA, resulting in the condensating and shortening of the immensely long DNA chains that build the genome of multicellular organisms. The condensed form of the DNA is called chromatin. It is still a mystery, however, as to how the packaging is dissolved before the cell divisions take place so that the DNA can be replicated, and how the DNA is identically rewrapped. In some regions of chromatin, the packaging is so dense that gene transcription is no longer possible.

A large portion of the DNA of eukaryotes consists of regions that are not transcribed and are hence non-coding. The coding regions that are ultimately translated into proteins can be recognized by the absence of stop codons. In contrast, non-coding regions are frequently interspersed with stop codons. Such non-coding regions are particularly long when they lie between genes. Regions that interrupt the coding sections within a gene are called introns, and those that are translated into proteins are called exons. Some genes of higher organisms have many introns so that their exons are distributed over very large DNA regions.

Genes are transcribed in the nucleus. The transcription process encompasses the entire length of the gene, producing an RNA transcript that includes the introns. Then the RNA molecules are processed by enzymes that can recognize the boundaries between introns and exons. In this manner, the introns are removed, the mRNA is transported to the cytoplasm, and, lastly, translated on the ribosomes (Figure 16).

Single-cell organisms, such as yeast, have only short regions between the genes. In contrast, mammals have stretches that are often much longer than the transcribed regions themselves. These areas between the genes contain control regions that bind to the transcription factors for guiding gene activity. In addition to the promoter region where the transcription starts, there are also enhancer regions that have an effect on gene transcription over long distances. Sometimes enhancers can be found in the introns. Transcription factors bind to these control regions and determine whether genes will be transcribed in the area under their influence.

The transcription of genes that fulfill general cell functions required in all cells is often regulated in a simple fashion. Other genes, especially those guiding development, have more complex control regions and can be bound by many different transcription factors with both activating or inhibiting effect.

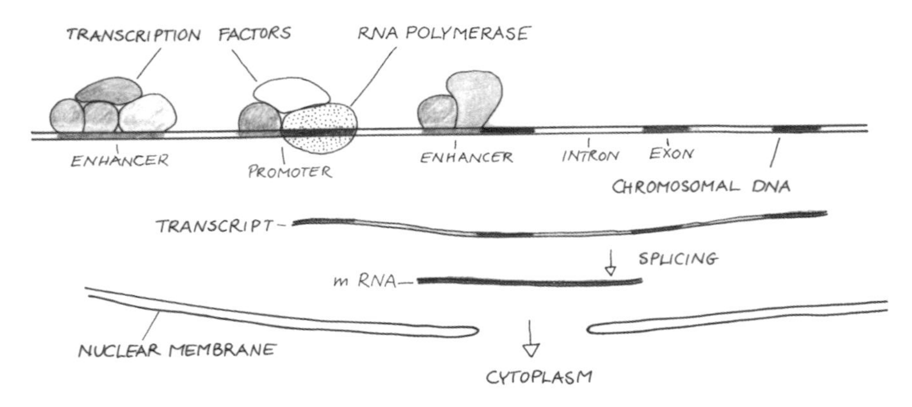

Figure 16. *Gene Structure of Eukaryotes.* Numerous proteins that work either as repressors or activators guide transcription. They bind not only to the promoter near the starting point of the transcription, but also to regions called enhancers that may be far away from the starting point. Initially, the resulting RNA also contains copies of gene regions that will not be integrated in the mRNA. These are called introns. After transcription, the introns will be spliced out so that the mature mRNA only consists of copies of the exons of genes. The mRNA is transported into the cytoplasm and used as the matrix for the protein synthesis.

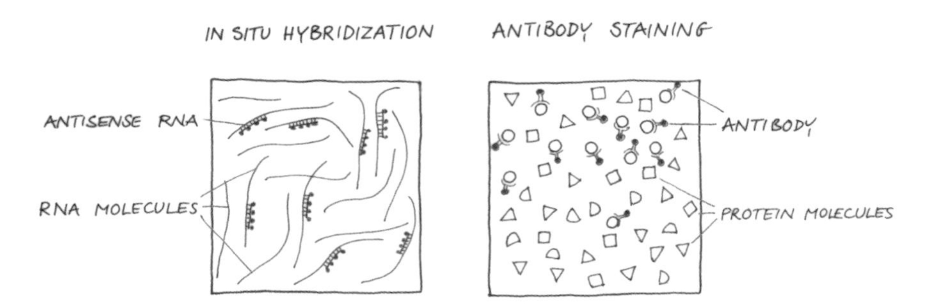

Figure 17. *In situ Hybridization/Antibody Staining.* The distribution of specific mRNAs or proteins can be visualized in a tissue sample by using color-labeled RNA molecules or antibodies. Labeled antisense RNA is complementary and binds to the mRNA of a given gene, thus highlighting those parts of an embryo where the gene is transcribed. Antibodies recognize and bind to specific proteins. Again this allows for staining the sites in an embryo where the protein is present. Refer to Figures 26, 27, 31, and 50 for examples of the application of this technique.

Distribution of RNA and Proteins. Molecular probes make it possible to visualize gene products, mRNA, and proteins in the organism. In order to detect mRNA, a complementary RNA copy of the gene is produced in vitro. This RNA is then marked with a dye and allowed to form hybrids with its endogenous mRNA. In cells where the gene's mRNA is present, the marked RNA is bound by hybridization, whereas in cells where the gene's mRNA is absent, the RNA probe is washed out. Molecular probes also make it possible to similarly see the distribution of proteins by using antibodies marked with a dye (Figure 17).

Transgenic Animals. Drosophila was one of the first higher organisms into which foreign genes could be introduced, through a process known as

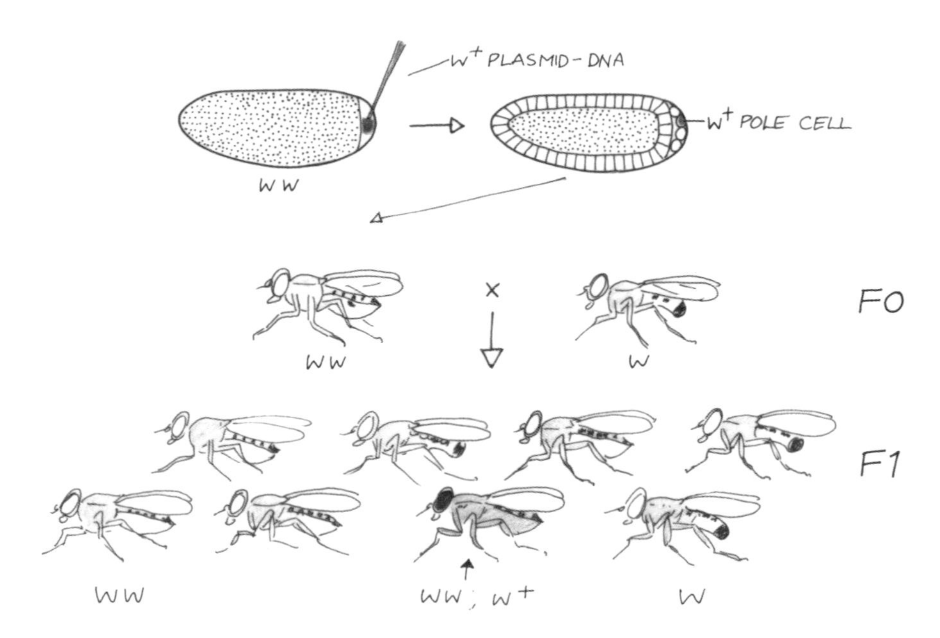

Figure 18. *Transgenic Flies.* A solution containing the DNA to be introduced into the fly genome is injected into the pole plasma. The plasmid also contains a copy of the white-gene, indicated as *w+*. The experiment is carried out with white mutant flies, such that when the plasmid is incorporated into the genome of a pole cell, a fly emerges (F0) that has some w+ germ cells (red). Among the progeny of these flies (F1) are so called "transgenic" flies that carry the extra gene in all the cells of their body and germ line. They are recognized by their red eyes and will transmit the extra gene to subsequent generations. The transgene has entered their germ line, which is why the technique is known as germ line transformation.

transformation. Animals carrying an additional gene are called transgenic. The creation of transgenic flies requires that a solution of plasmid DNA containing copies of the isolated gene synthesized in bacteria be injected into the cytoplasm at the posterior end of the egg. Later, this cytoplasm is encased within the pole cells from which the germ line develops, resulting in some of the injected flies carrying the additional gene in the genome of some of their gametes. These gametes produce offspring carrying the additional gene in all cells of the body (Figure 18).

Marker genes that change visible features such as eye color are used to recognize the transgenic offspring. This method can, for instance, rescue mutants by transforming them with the intact isolated gene. Hence, it can be tested whether the gene in question was correctly isolated, and whether it contains all of what is necessary for the normal function of that gene. The genes used for transgenesis can also originate in other organisms or can be put together from sections of various other genes. Transgenic animals allow us to learn a lot about the normal function of a gene as well as its relationship to other genes within an organism. Transgenic animals also allow the study of what happens when a gene is overactivated or activated at the wrong place or time. These are only a few examples of the many powerful applications of transgenic animals.

CHAPTER IV

Development and Genetics

THE EXCITING PROGRESS IN CLASSICAL AND MOLECULAR GENETICS HAD THE effect of pushing into the background the study of embryonic development. Indeed, after Hans Spemann's great experiments in the first decades of the 1900s, there was barely any progress in this field until the 1970s. Even though the phenomenon of embryonic induction, the manipulation of neighboring regions in the embryo through hypothetical organizing signals, had been described in various contexts, the molecules involved had still resisted detection. However given the difficulty of the embryological experiments and the lack of hard facts, there were significant advancements in the development of theoretical concepts. These concepts proved to be quite interesting because they helped in imagining how embryonic induction might actually work.

Based on observations from regenerating organisms such as the fresh water polyp Hydra, South African biologist Lewis Wolpert (1929–) at University College London, UK, pointed out that the position of cells within the developing embryo determines their fate. His concept of positional information proposed that the distance from an organizer can be measured by the cell as the concentration of a substance that is spreading away from the organizer. Embryonic cells require a certain concentration of that substance in order to reach a particular state or quality: those cells exposed to concentrations of the substance above a threshold concentration will be different from those at lower concentration. This will then subdivide the cells into two zones. A substance that elicits different responses at different concentrations is called morphogen, and the gradual change of concentration is called gradient. Other cell fates may require different threshold values of the same substance. In this manner, one morphogenetic gradient can determine a series of several states (Figure 19).

However, gradient models, even though they can explain an increase of spatial complexity in a rather simple manner, were not very popular among biologists. This is because for a long while the postulated morphogens could not be identified using the conventional methods of biochemistry and embryology.

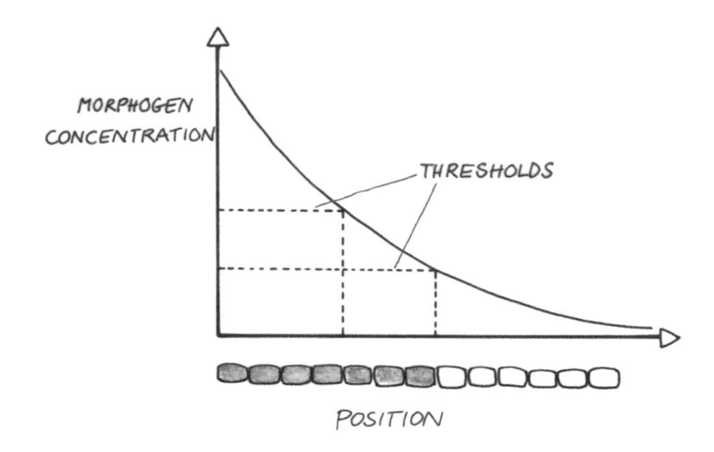

Figure 19. *Morphogenetic Gradients.* A morphogen is a substance that causes cells to adopt different properties or give rise to different structures depending on its concentration. The minimal concentration necessary for a certain structure is called threshold. In the example given here, the differentiation of gray cells takes a higher threshold concentration than that for the differentiation of red cells.

How do you isolate the factors guiding the major steps of embryonic development? What kind of molecule mediates the activity of Hans Spemann's organizer? What is the something described by Theodor Boveri that decreases or increases inside the egg from top to bottom? Do morphogens really exist, and if so, what do they look like?

Attempts to isolate such factors using biochemical enrichment methods failed, mainly because these molecules are usually present only in trace amounts. Furthermore, the assays used to test whether these factors were present in a biochemical extract were often very unreliable. However, as it was likely that these factors were proteins, there would have to be genes encoding them, and they could be identified using genetic approaches. For example, we would expect embryos mutant for the gene for Spemann's organizer to lack a body axis. Furthermore, in bacteria, systematic searches for mutations proved to be particularly powerful not only in dissecting metabolic pathways, but also crucial in finding the most important regulators of transcription.

Influenced by these successes in bacterial genetics, similar approaches were undertaken in the search for mutations in genes that have guiding functions in the formation of multicellular organisms. Thus, a new field of research, called developmental genetics, emerged in the 1970s. Among the results was the discovery of morphogens as well as the signaling molecules during induction and revelations about their molecular nature.

1. *Model Organisms*

Genetic research needs to focus on one species, as mutants in one animal species cannot be crossbred with those of another one. Therefore, the subject of research must be chosen carefully. Frogs are not suitable for genetic research because the adult frogs and their offspring need a lot of space, and because the generation time is 2 years. Sea urchins are not suitable because they cannot be bred in the laboratory. In the 1970s, at Cambridge, South African biologist Sydney Brenner (1927–) chose the worm *Caenorhabditis elegans* as a model organism because of its simple body organization and ease of genetic experimentation. When fully grown, this little worm has a constant number of 959 somatic cells that develop in a strict succession of divisions, all of which have now become well understood. Also, subtle details can be studied during the differentiation of individual cells and their interaction. This worm turned out to be an excellent choice of study in offering various advantages to researchers.

The fly *Drosophila* also seemed a highly suitable model organism for a systematic search for genes that specifically influence embryonic development. At that time, the embryology of the fly compared to the frog was much less understood; however, the fly also proved to be an excellent choice for genetic studies of form building processes. Developmental studies first concentrated on the formation of the larva because the larva and not the adult fly is the immediate product of embryonic development. Mutations of developmental genes would most likely result in phenotypes that affected the larval organization.

The outside skeleton of the larva, known as the cuticle, has a variety of easily visible structures that allow the overall integrity of the larvae to be easily assessed. The bottom or *ventral* side is marked by bands of little hooks that help the larva crawl, whereas the back or *dorsal* side displays hairs that are arranged in a particular pattern. Likewise, the thoracic segments at the front can easily be distinguished from the abdominal segments in the rear. Numerous sense organs and mouthparts develop at the head while the nonsegmented hind end also shows characteristic structures.

2. *The Development of Drosophila*

Egg Formation. Considering the size of the adult fly, the egg is rather big, about 0.5 millimeters long. The front, rear, top, and bottom of a fly egg are easily distinguishable by the shape of the eggshell. The egg is produced within the mother from two types of cells. The first are germ-line cells that give rise to the egg cell and the supporting nurse cells. The second are somatic follicle cells that will later form the eggshell. Nurse cells and follicle cells transport proteins and fatty substances into the growing egg cell, which nourishes the embryo until it hatches. After copulation, the mother fly keeps the sperm in a semen bag to fertilize the egg before laying. But despite its huge size, it is virtually impossible to distinguish any structures within the egg cell except for the clear region at the posterior pole, called pole plasm. This pole plasm will be incorporated into the first embryonic cells, which will then later give rise to the germ line.

Embryonic Development. The egg nucleus can be found in the top frontal part of the freshly laid egg. During the first hour after egg laying the nuclei divide, but the cytoplasm does not. In several quick divisions, many nuclei form and migrate to the surface to divide several more times. At this stage, the embryo consists of one very large cell with many nuclei and is referred to as syncytial blastoderm. Unhindered by cell membranes, its nutrients and other substances can spread easily. Three hours after fertilization, membranes are formed simultaneously between the 6000 nuclei. This stage is called the cellular blastoderm, at which all the cells still look the same.

During the process of gastrulation, groups of cells migrate and fold into the embryo to build an inner cell layer or *endoderm*, which will later form the intestines and stomach; they also form the intermediate cell layer or *mesoderm*, which will later form musculature, heart, blood, and other inner organs. The outer cell layer or *ectoderm* remains at the outside and will later develop into the nervous system and the skin. During gastrulation, a furrow develops along the ventral side of the embryo. Cells at the front and back end of the embryo, which will develop into the intestines, also begin to fold in. On its dorsal side, the embryo stretches forward. In the stretched state, the embryonic cells divide further, the nervous system is established, and the organization of organs and segments begins to appear (Figure 20). Further movement of cells to their final positions characterizes later development. Finally, differentiation takes place with cells specializing according to their functions within the individual organs. After approximately 24 hours, the larva hatches.

The heart of the larva develops on the dorsal side. There are no arteries or veins, so the colorless blood that carries nutrients and hormones circulates freely around the inner organs of the larva. Oxygen is not carried by the blood but rather enters the larva through an extensive network of air-filled pipes, so-called tracheae, which is connected to the outside in the front and back.

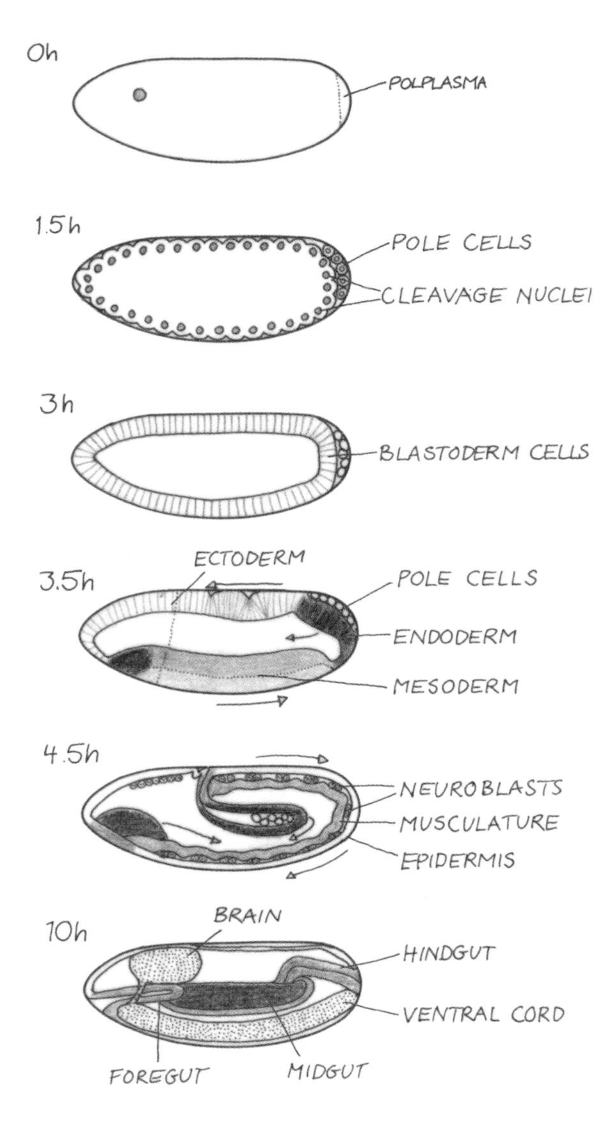

Figure 20. *Embryonic Development of Drosophila.* During cleavage, a syncytial blastoderm develops (at 1.5 hours), in which all the nuclei share the same cytoplasm. By 3 hours after fertilization, membranes form between the nuclei, giving rise to the cellular blastoderm. During the process of gastrulation (at 3.5 hours), groups of cells giving rise to the endoderm (red) and the mesoderm (gray) migrate inside. Further development is accompanied by a stretching of the embryo (at 4.5 hours), which is reversed later (arrows). The primordia of the digestive tract illustrated here in red are now connected in the center and the ventral nerve cord (stippled) has developed from neuroblasts. The anterior is to the left and ventral to the bottom. The numbers on the left refer to hours after fertilization.

The central nervous system of the larva consists of a brain and a rope-ladder-shaped ventral nerve chord, both of which develop from individual cells called neuroblasts. Neuroblasts separate from the rest of the ectoderm cells in a characteristic pattern as they move inside and organize into knots and strings of nerves.

Imaginal Discs. As mentioned earlier, the larva basically grows through an increase in cell size rather than through cell division. The larva sheds its skin twice before it finally pupates. Compared to the larvae of other insects, for example, caterpillars, fly larvae have a relatively simple structure: they have no legs and their head is very small. This facilitates shedding the skin, which is a necessary function because the outside cuticle coat limits expansion of the body. Inside a pupa, the larva metamorphoses into the adult fly, which, after 3 days, hatches through the front end of the pupa. The hormone ecdysone triggers metamorphosis, and the cuticular structures differentiate in the pupa, resulting in the development of bristles, hairs, eyes, sexual organs, and wings. Just before hatching they fold to the outside and connect to each other without leaving traces of seams. The body of the fly then emerges.

During metamorphosis the outer structures of the fly do not develop directly from those of the larva but rather from undifferentiated cells, called imaginal discs. Many such imaginal discs fit together like a puzzle to form the fly. The imaginal discs come in pairs, one disc for the right side of the body and one for the left. There are three pairs for the six legs, two pairs for the wings and the thorax, two for the head with eyes and antenna, and a few smaller ones for parts of the abdomen and the head. Every imaginal disc grows from a small group of 3 to 10 precursor cells that have been set aside in the embryo. These cells divide regularly such that the mature imaginal discs consist of layers of up to 40,000 cells (Figure 21).

Fate Map. By the labeling of specific groups of cells in early embryos, it is possible to determine the origin of the various body structures of the larva or the adult fly. At first, the fate map of the larva is two-dimensional with a clear system of coordinates forming two axes perpendicular to each other: front versus rear end, or anterior versus posterior; as well as top versus bottom, or dorsal versus ventral. The cuticle develops from a large region comprising about half the length of the egg. Each segment develops from a row of 3 to 4 cells. Within these rows, small groups of cells in the frontal segments contain the precursor cells for the imaginal discs. The cells that will ultimately form the inner organs run along the ventral side. Furthermore, large

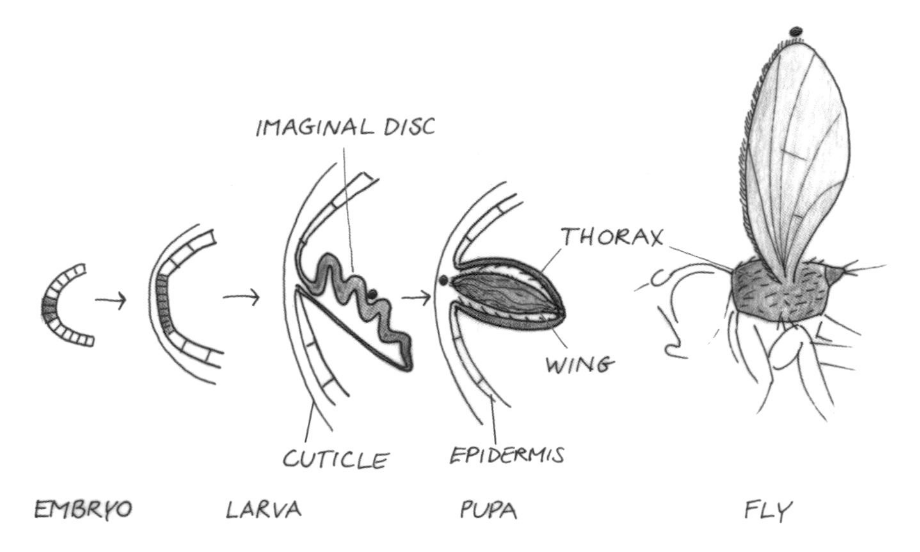

Figure 21. *Imaginal Discs.* The imaginal discs come from cells in the ectoderm and will form all of the adult fly's external structures. They develop into sac-like folds, shown here in cross section, and differentiate during metamorphosis under the influence of the hormone ecdyson. They then unfold to give rise to the fly's adult structures. The external parts of an adult fly are constructed like a mosaic from several imaginal discs. Illustrated above is the wing disc that forms the thorax as well as the wing blade shown in gray in the center. The red dot marks the tip of the wing.

anterior or posterior areas will move inside the embryo to form the brain and the intestines (Figure 22).

3. The Search for Genes that Control Development

How does the fate map develop? Is it already present in the egg as assumed by the theory of preformation. Or does it develop from scratch, as our logic tells us? Furthermore, how much information for the pattern-forming processes is already present in the fertilized egg, and how much develops later?

All substances that guide development can be traced back to genes. As mentioned previously, *Drosophila* has about 5000 genes that are necessary for the normal development and survival of the fly. Most of these genes produce proteins that are needed for general cell processes, such as cell growth and division, resulting in the formation of new cellular building blocks. This is not true, however, for a small group of genes that have the special function

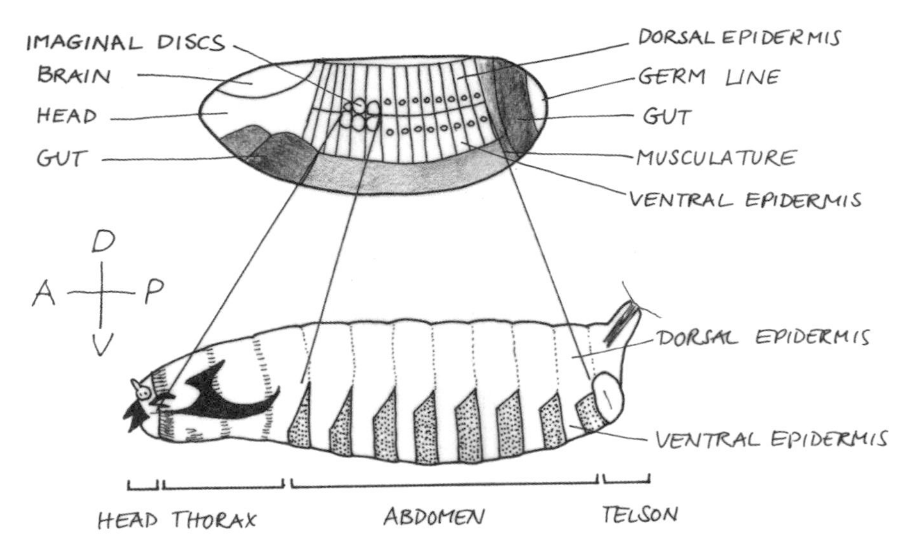

Figure 22. *Fate Map of the Drosophila Embryo.* The fate map shows which regions of the blastoderm embryo (upper panel) will give rise to certain regions of the larva (lower panel). The segmented area of thorax and abdomen that covers the larva's body from front to back develops from a large region in the center of the embryo. Within this area, each segment consists of a small stripe that is only three cells wide. Most of the other regions will move inside the embryo during gastrulation and develop into inner organs such as the digestive tract (red) and the musculature (gray). *A* stands for anterior, *P* for posterior, *D* for dorsal, and *V* for ventral.

to guide development during pattern formation of the larva or the adult fly. These genes are usually active only at certain times and in certain regions of the embryo, resulting in a specific spatial organization. Mutants of these genes show a visible phenotype: the larva's structures are either in the wrong place, misshapen, or missing altogether.

The early developmental stages of the larva are determined by factors already present in the freshly laid egg, all of which are synthesized from the mother's genes and thus depend on her genotype, so-called maternal factors. For example, mutant females with a defective gene for one such factor produce eggs that can only develop into misshapen larvae. The phenotype of such maternal mutations only shows in the offspring and is not influenced by the genotype of the father (Figure 23). Later, starting with the development of the blastoderm, the embryonic genome becomes active and the phenotype from this stage on is determined directly by the genotype of the embryo; that is, the combination of genes it inherited from both mother and father. Mutants of

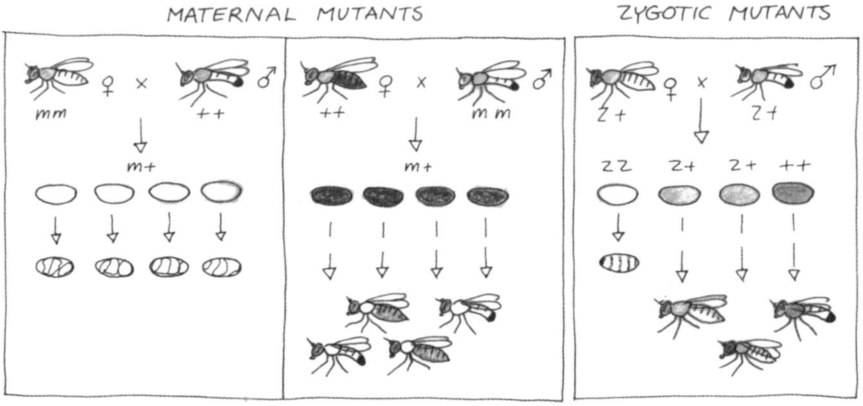

Figure 23. *Maternal and Zygotic Mutants.* Females that are homozygous for any given maternal mutation (mm) produce eggs that lack something (red) needed for a normal development, even if the paternal genome contributes the normal allele (left and middle panels). This means that it is the mother's genotype that decides whether the mutant phenotype will show in the embryos or not, as the embryo's genotype is identical in both crosses. Zygotic mutants (z) only show the phenotype in homozygous mutant embryos (zz) that are a result of a cross between two heterozygous (z+) parents.

those genes are called zygotic, as the genotype of the zygote determines whether a phenotype shows or not (Figure 23). Some cases of such mutants had been found in the collection of *Drosophila* laboratory strains. But since the detection of such mutants requires careful study of the larva, the majority was discovered in specific laboratory experiments.

To find mutants, male flies are fed with a chemical that increases the error rate of their DNA replication. Originating from one mutant sperm, families are inbred in a way that homozygous individuals develop after several generations. To find mutants in most, if not all genes, many inbred families must be examined. The precise number is determined by the success rate of the chemical. The mutant's cuticula survives the deterioration of tissue, even after death, and the cuticular pattern is suitable for assessing the effects of a mutation because it covers the entire body and reveals subtle changes in shape and form.

In systematic searches maternal and zygotic genes were found, mutations that lead to the development of larvae with defective structures or a defective body organization. Approximately 40 maternal and 120 zygotic genes that specifically determine the larval cuticular pattern have been discovered in this way. Interestingly, they comprise only a very small fraction of the fly's genome.

4. The Logic of Genes

A mutant defines a gene. The phenotype of a mutant animal with only one defective gene, while all others are normal, can provide much information. Which structures are affected by the mutation and which ones are not? How early in development does the mutation show and what are its effects? In the study of pattern formation, the *Drosophila* mutant collection is of particular value because it contains many if not all genes controlling development. We then end up holding most pieces of the puzzle in our hand and can start putting them together.

What do the mutants look like and what is the logic behind them? Contrary to what one might expect, it is very rare that one gene determines only a single structure. In other words, there is not a single gene for each body segment, nor each leg or organ or bristle, and so on. The logic of genes does not follow morphologically visible subdivisions. The fate map, therefore, cannot be explained by genetic products that represent certain structures in a certain position, as would be expected if the theory of preformation was valid. The number of genes, and especially the number of phenotypes, is simply much too small for that.

The phenotypes of various mutants are often similar. In fact, several mutants commonly share the same or very similar phenotype. Sometimes this is due to several different alleles of the same gene, but more often this is due to mutants in different genes that must have similar developmental functions. This very similarity allows for the classification of genes into specific groups. It stands to reason then that the products of all genes within one group will be relevant for a common process. Perhaps this is similar to the processes of cell metabolism, whereby a chain of biochemical reactions produced by individual enzymes leads to the synthesis of a certain product. Likewise, perhaps a mutation in individual genes of the chain leads to a similar phenotype, such as the lack of a certain body region during the process of pattern formation.

Mutations of maternal genes lead to particularly dramatic changes in the shape of the larva. This makes perfect sense as maternal genes affect the earliest functions of pattern formation. For example, embryos from mutant female flies lack large body regions. Interestingly, there are only a small number of possible defects: either the changes are in the pattern along the anterior–posterior axis, or in the pattern along the dorsal–ventral axis. This tells us that both axes are formed independently from each other. Most mutants of the dorsal–ventral axis lack a ventral side such that the dorsal structures of the embryo continue all the way around its circumference. The subdivision of the anterior–posterior axis, however, is not affected in these mutants. Genes influencing the anterior–posterior axis can be divided into three groups: one responsible for the front segments, one for the segments of the rear, and one for the two non-segmented ends at the egg poles (Figure 24). Thus, the maternal developmental program is divided into four independent processes. Each process determines only a part of the normal pattern.

There is more variety among the phenotypes of zygotic genes. The missing regions are usually smaller, and the defect is visible only after gastrulation. Within this class, there are also groups with similar or identical phenotypes.

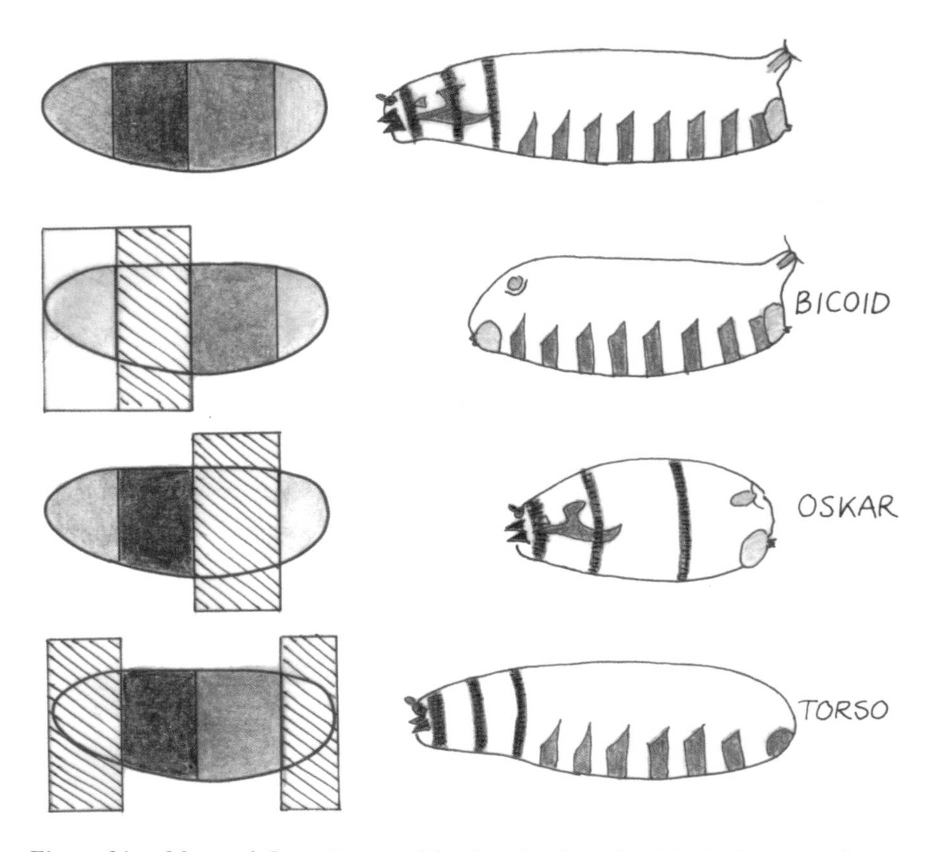

Figure 24. *Maternal Gene Groups of the Anterior-Posterior Axis.* Embryos produced by maternal mutant females lack characteristic parts of the segment pattern. Bicoid embryos lack all structures that would normally result from the front half of the egg, oskar embryos lack the abdomen, and in the case of torso both the very front and the very back are missing. There are several genes with phenotypes like oskar or torso. The products of the genes of such a group are often active during the same process, one after the other. A fourth maternal gene group determines the dorsal-ventral axis (not illustrated here).

One of these zygotic gene groups influences pattern formation along the dorsal–ventral axis of the embryo in a way similar to the corresponding group of maternal genes. The missing structures, however, are smaller and affect only parts of the pattern.

In several mutants, the number of segments is reduced. In addition to the maternal genes affecting large body regions, there are three groups of zygotic genes that are involved in segmentation. The phenotypes of the first group, known as gap-genes, are similar to those of the maternal genes in that the embryos lack entire regions of the body. However, these missing regions are smaller than those missing from maternal mutants. For example, the phenotype of the maternal gene oskar (Figure 24) is very similar to that of the zygotic gene knirps (Figure 25). The other two groups of zygotic genes affect several segments: mutants of the "pair-rule" class of genes lack corresponding

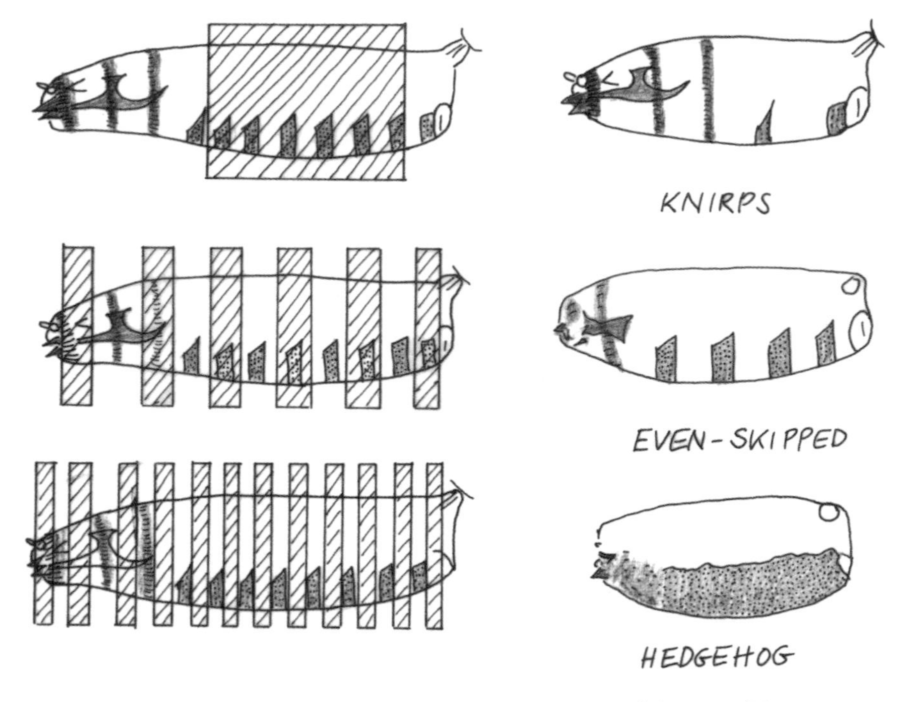

Figure 25. *Zygotic Segmentation Genes.* Knirps (top panel) is one of the gap-gene mutants which cause the lack of large continuous regions. In the case of knirps, almost the entire abdomen is missing. Pair-rule mutants like even-skipped (middle panel) lack regions in every second segment. Mutants of segment-polarity-genes like hedgehog (lower panel) lack parts of every segment – in the case of hedgehog the naked cuticle.

structures in every other segment; and the "segment-polarity"-genes affect every single segment. This suggests that the embryo is subdivided into steps from larger to smaller units. In a first step, large regions are defined coarsely by an interplay between the maternal genes and the gap-genes. Subsequently, a periodic pattern of double segments appears, driven by the pair-rule-genes. This pattern is then further refined by splitting the double segments into single segments.

The phenotypes point to a hierarchy of gene function: those genes that are active early in the process control the effects of those genes that are active later on. Indeed, as we shall see in the next chapter, molecular analysis revealed that many of the segmentation genes encode transcription factors that act as morphogens. This means that the genes that guide development do so by regulating the activity of other genes.

CHAPTER V

Molecular Prepatterns

DROSOPHILA EMBRYOS ARE SPECIAL BECAUSE DURING EARLY STAGES OF development only the nuclei of the zygote divide and no cells are formed yet (Figure 20). The nuclei are arranged along the periphery of the egg and share the same cytoplasm such that proteins can spread between them. In other words, the early *Drosophila* embryo is one giant cell containing many nuclei. Maternal and zygotic factors controlling early development can disperse freely in this cell.

Before fertilization of the egg, the four maternal gene groups provide four cues that are positioned at the anterior pole, the posterior pole, and on the ventral side. From there, the cues spread and organize the pattern of the entire embryo (Figure 26). These maternal factors regulate the activity of the first zygotic genes, which encode short-lived transcription factors that either activate or repress other genes. They can diffuse away from the nuclei in which they are originally synthesized and bind to specific enhancer regions of other genes, thus guiding their activity. Through an interplay of mutual activation and repression, more and more complex molecular patterns emerge. Unless visualized by special techniques, these patterns of gene activity are invisible, and hence referred to as prepatterns. At the end of the blastoderm stage, the final prepattern triggers the first changes in cell shape, which then lead to the visible structures of the embryo.

The development of molecular prepatterns relies on two important principles: first, different concentrations of a factor elicit different responses; and second, several factors operate in combination. As described in Chapter IV (see Figure 19), morphogenetic gradients with different thresholds can create zones of different molecular identity. In addition, the combination of two or more factors may create a third, new identity. These two principles are crucial for embryonic development because they allow the progression from simple patterns to more complex ones.

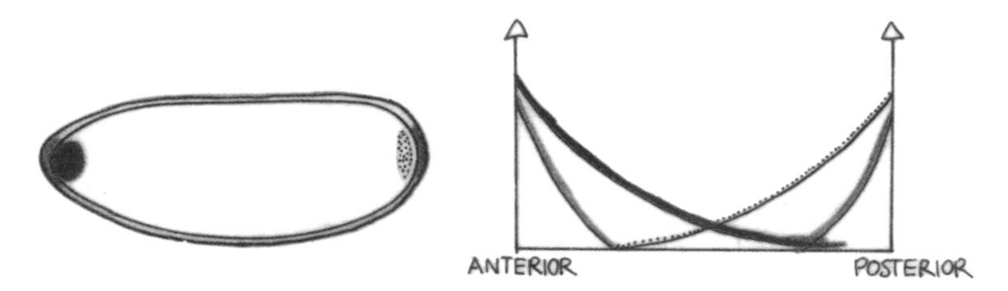

Figure 26. *Maternal Prepattern of the Anterior-Posterior Axis of Drosophila.* Some gene products of the maternal genes are already localized in the egg (left panel). They act as signals guiding the development of the embryo from the periphery. Bicoid mRNA (red) is located in the front and oskar-mRNA (stippled) is located at the back. The torso-protein is a receptor protein that is activated in the membrane at both of the egg's poles (black). Gradients develop starting at these signals (right panel). The simplest case is the gradient (red) of the bicoid protein at the anterior. Torso activation results in two short range gradients at both poles (dark gray).

Later in development, the cells of the *Drosophila* embryo form as the membranes arise between the nuclei. At this stage, factors can no longer simply diffuse between nuclei. Thus, the cells start secreting signaling molecules that spread through the extracellular space and influence the neighboring cells via signal transduction pathways. These pathways allow the signal to cross the cell membrane of the receiving cell.

1. Gradients

A gradient is a gradual change of a quality over a distance, for instance, the decrease in temperature with increasing height on a mountain. This change in quality can trigger different outcomes. In our example, belts of various types of mountain flora emerge as a result of the temperature gradient along the mountain. There are also density gradients, color gradients, and concentration gradients of morphogens—such gradients are vital to an embryo's development.

One of the first cases of a morphogenetic gradient for which an outcome could be clearly shown to depend on the concentration, was the "bicoid" gradient in *Drosophila*. The bicoid-gene belongs to a group of maternal genes that are responsible for the formation of the anterior pattern of the embryo (Figure 24). If the bicoid-gene is defective, the embryos from the mutant mother will lack all structures that are usually formed from the front half of the embryo: the head and the thorax. The bicoid mRNA is anchored at the anterior pole of the freshly laid egg. During the first hours after fertilization, the bicoid protein is produced through translation of this localized mRNA. The bicoid protein diffuses toward the posterior pole of the egg, creating a gradient across the egg with the highest concentration at the anterior and decreasing toward the posterior (figure 26). The bicoid protein is a transcription factor that contains a homeodomain that binds to the promoters of various segmentation genes. Among them is the promoter of the hunchback-gene, a member of the gap-genes (see page 55).

Through the action of bicoid, the hunchback-gene is transcribed in the nuclei of the anterior half of the egg. If the bicoid protein is missing, then the hunchback-gene is not transcribed. If the gradient of the bicoid protein is experimentally changed and the bicoid protein concentration is higher or lower than normal, the gradient becomes steeper and higher, or lower and more shallow, respectively. Consequently, the point at which the bicoid protein concentration reaches the threshold for the transcription of hunchback moves forward or backward, respectively (Figure 27). These experiments by

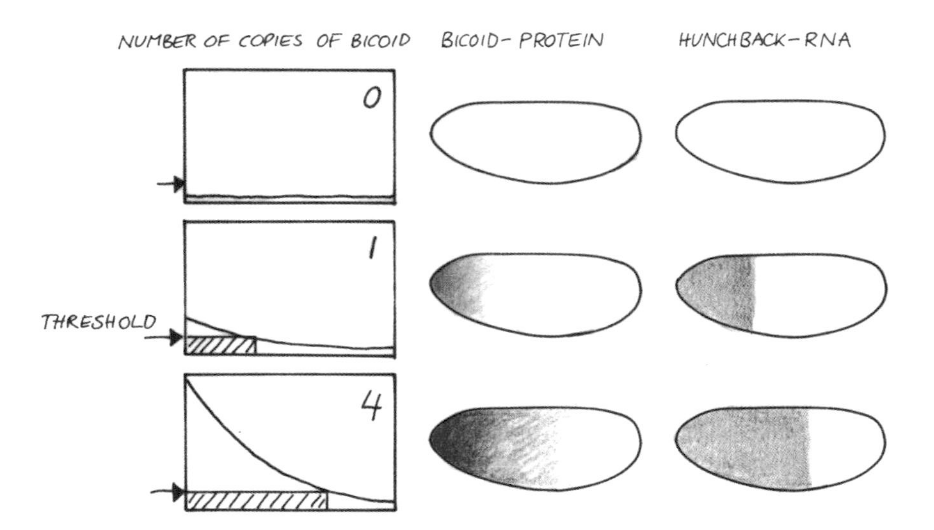

Figure 27. *Concentration Dependency of Hunchback Activation.* The bicoid gradient can be modified by manipulating the amount of bicoid mRNA located in the egg. Mutants lacking the bicoid-gene altogether (top panel) obviously form no gradient. Heterozygous females (middle panel) only have one intact bicoid-gene copy, whereas transgenic females with two additional bicoid-gene copies (lower panel) produce eggs with four times the amount of RNA. The gradient, visualized as bicoid protein distribution in the egg (middle column), changes accordingly, moving the position of the domain in which hunchback mRNA is transcribed (right column).

German biologist Wolfgang Driever and myself demonstrated that a minimal concentration of bicoid protein, normally present in the anterior half of the egg, is sufficient and necessary to activate the hunchback-gene. Another gene whose promoter is bound less effectively by the bicoid protein will be transcribed at higher bicoid protein concentrations. These higher concentrations are only present in the anterior-most region of the egg. With two genes responding to different thresholds, three different zones are defined: at the highest concentration, both genes are active; in a region with a medium concentration, only one is active; and finally, in a region with the lowest concentration, neither of the two is active (Figure 28).

This example shows that a gradient of one morphogen can subdivide the egg into several zones. The bicoid morphogen diffuses from the point of origin at the anterior pole of the egg where it is continuously produced from the

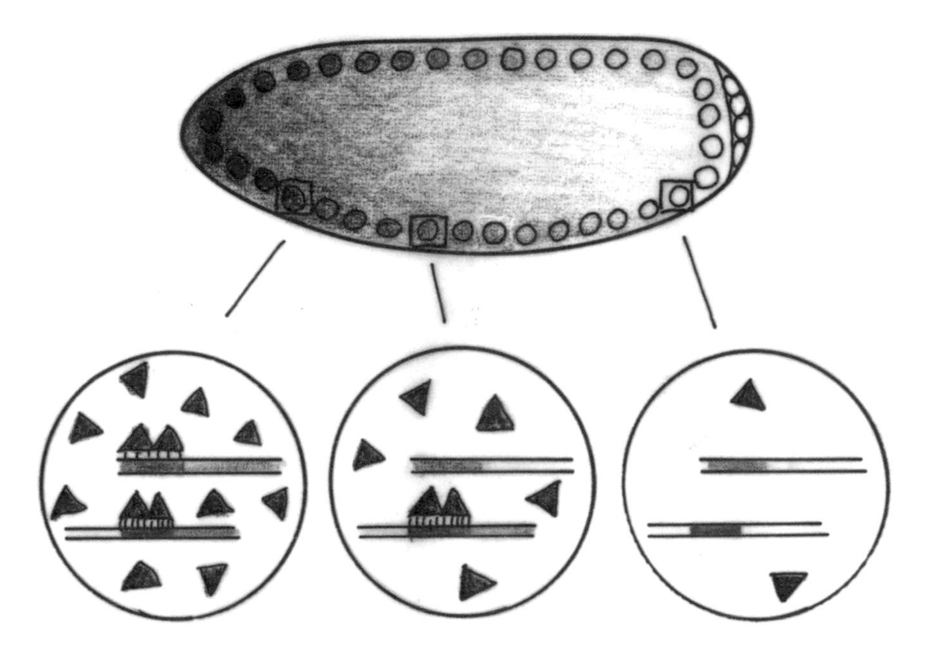

Figure 28. *Different Thresholds of Gene Activation.* A promoter that binds the bicoid protein with high affinity is shown here in black. It will be activated by concentrations above a relatively low concentration of bicoid protein, as are present between the front and the middle of the *Drosophila* egg. A second gene, shown here in gray, is also activated by bicoid yet only at higher concentrations (left circle). The two genes have different thresholds of activation by bicoid protein. At very low concentrations neither of the two genes is activated (right circle). Hence, the bicoid gradient can determine more than just one position within the egg.

anchored mRNA. The bicoid protein is unstable and decays rapidly. If the protein was stable, it would accumulate in the egg and a gradient would not form. The continuous decay of the bicoid protein counteracts its synthesis, generating a graded distribution in the egg.

The four maternal gene groups described in Chapter IV create gradients in the early embryo. One of them is the bicoid gradient, which reaches up to the posterior third of the embryo. Two additional gradients, from the torso-gene group, have shorter ranges: one originating from the anterior pole and the other from the posterior pole (Figure 26). An mRNA localized at the posterior pole, the oskar mRNA, is responsible for the formation of the pole plasm from which the germ cells emerge. A protein gradient starts from the pole plasm and spreads toward the anterior of the egg. This gradient controls the transcription of knirps, a gene required for the development of the abdomen. This explains the similarity of the oskar and knirps phenotypes (Figures 24 and 25). In general, the maternal gradients regulate the transcription of zygotic segmentation genes, the gap-genes, such as hunchback and knirps, along the anterior–posterior axis. Lastly, the subdivision of the embryo along the dorsal–ventral axis depends on only one maternal gradient, the "dorsal" gradient.

The Dorsal Gradient. A gradient of the protein product of the dorsal-gene covers the dorsal–ventral axis of the embryo, reaching its peak on the ventral side. The localized signal that triggers this gradient is not the mRNA of the gradient molecule, but rather it is provided by a protein named Toll, which is anchored in the membrane throughout the egg. Toll protein activity causes the dorsal protein, which is present in the cytoplasm, to enter the nuclei. Despite its broad distribution, Toll protein is only active on the ventral side. Consequently, dorsal protein enters the nuclei only on the ventral side of the embryo and remains in the cytoplasm on the dorsal side. This creates a gradient of nuclear dorsal protein (Figure 29). As the dorsal protein is a transcription factor, it can only function when it is in the nucleus where it controls the transcription of several genes. At high concentration, the dorsal protein activates twist, a gene required for the formation of the musculature. Lower concentrations of nuclear dorsal protein turn on different genes in the adjacent zones on the right and left sides of the embryo. The dorsal protein also represses the transcription of certain genes. Consequently, these genes are active only on the dorsal side of the embryo where dorsal protein is not in the nucleus. In this manner, the dorsal gradient divides the egg into four longitudinal zones. Some of these zones immediately result in cell shape changes such as the domain in which the twist protein is activated.

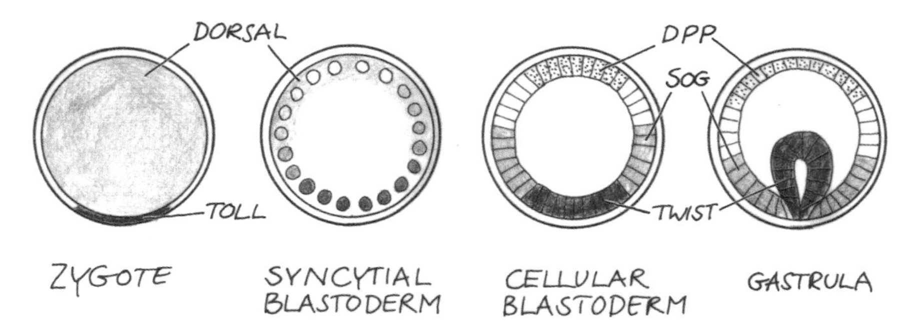

Figure 29. *The Dorsal Gradient.* The dorsal gradient develops when the initially evenly distributed dorsal protein, shown here in gray, gets transported into the nuclei on the ventral side, while the dorsal protein stays outside of the nuclei on the dorsal side (second panel). This nuclear localization is a result of a signal transduction process in which Toll functions as the receptor in the egg membrane. It is activated by an outside signal at the ventral side (shown in the leftmost panel in red). The gradient of dorsal, which is a transcription factor, works similarly to bicoid in that varying concentrations activate or repress different genes. This results in the generation of twist protein (red) on the ventral side, sog protein (gray) next to it, and Dpp protein (stippled) on the dorsal side.

These cells invaginate into the embryo to form the mesoderm. The two adjacent zones regulate the differentiation of the ectoderm, which will form the nervous system and skin. Finally, the zone at the dorsal side will form the fine hair pattern that covers the back of the larva (Figure 29).

The subdivision of the anterior–posterior axis is more complicated. It is created by a series of transient molecular prepatterns eventually developing into the visible pattern of the segmented embryo.

2. *Combinations*

In theory, it is possible that the different concentrations of a single morphogen determine a whole series of different developmental states. But in most cases, other factors are active as well and overlap with the morphogen function. For example, combinations of several transcription factors may be present in one nucleus at the same time. These combinations can create new states. If, for example, factor "a" determines the developmental state "A" and factor "b" the state "B," the presence of both factors at the same time may create a new state "C." Additional diversity can arise when, for instance, an activator and a repressor compete by binding simultaneously to the promoter of the same gene. By the same token, a single factor alone may act as an activator, but when combined with a second factor, it may become a repressor. Thus, whether a gene will be turned on or off is determined by a combination of these various mechanisms.

Gap-Genes. The gradients of maternal morphogens regulate the transcription of the gap-genes along the anterior–posterior axis. There are at least six gap-genes and all are transcribed in one or two broad stripes. The position of these stripes is initially determined by the maternal gradients. Later, these stripes of gap-gene expression are modified further and refined by interactions between the gap-genes themselves. This is because each gap-gene encodes a transcription factor that can activate or repress the transcription of other gap-genes. The gap-gene proteins can diffuse in the cytoplasm and form short-range concentration gradients. These gradients appear as bell-shaped curves (Figure 30, middle panel). The domain in which a gap-gene is transcribed is where its protein is most concentrated. At these high concentrations, the protein acts as a repressor of other gap-genes. As the concentration decreases around the margins of the domain, the protein concentration falls below a threshold value for its repressor activity, allowing the activation of one or more other gap-genes. This process creates adjacent peaks of gap-gene activity, which cover the egg sequentially from anterior to posterior.

After activation by the maternal gradients, the gap-genes create a simple molecular pattern of several broad stripes. Initially, these stripes are not well defined and change shape as the gap-gene proteins diffuse away from their point of origin and begin to influence each other. During this process, whether or not a given gene will be activated depends on both the concentration and the particular combination of factors that bind to the control region of a gene. In addition, transcription factors, such as the gap-genes themselves, may also stimulate their own transcription in a way that leads to a sharpening of the initially diffuse borders. A hypothetical example (Figure 30, top three panels) illustrates this mode of action.

Periodicity. The first zygotic prepattern (Figure 30, middle panel) created by the gap-genes is far more complex and contains more elements than the maternal morphogen distribution (Figure 30, top panel).Yet, one essential aspect of the fly's body plan is still missing from this prepattern, namely the subdivision into repetitive units such as the periodicity of the segments. The first periodic prepattern is formed when the gap-genes activate the so-called pair-rule-genes. Each of these pair-rule-genes is transcribed in seven regular

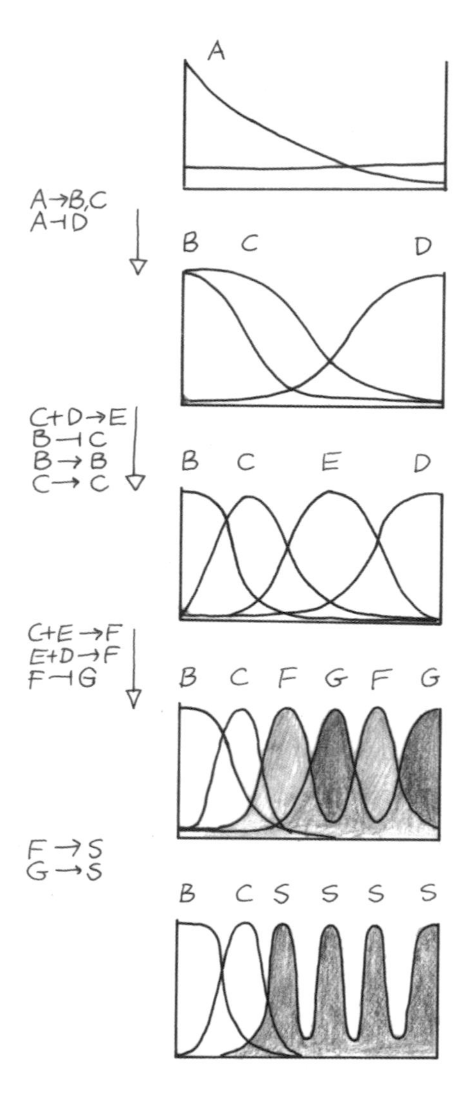

Figure 30. *The Development of Molecular Prepatterns.* The gap-genes and pair-rule genes encode transcription factors. In the early *Drosophila* embryo, these proteins can diffuse freely, unhindered by cell membranes. They control the transcription of other genes. This results in a series of more and more complex molecular prepatterns. They finally determine the periodic pattern that precedes segmentation of the embryo. This figure shows a simplified hypothetical series of such prepatterns. The horizontal axis shows the concentration of the protein and the vertical axis position in the embryo. The step-by-step transformation of these patterns is the result of a complex interplay of several factors influencing the same nucleus as activators, repressors, and self-enhancers. The symbol → signifies activation, and … signifies repression.

stripes, one stripe per two future segments. This explains why embryos with mutations in a given pair-rule-gene show defects in every other segment (see Figure 25). For instance, the mutant of the even-skipped-gene has defects in the even-numbered segments, while fushi-tarazu mutants lack odd-numbered segments.

Surprisingly, the seven stripes of the different pair-rule-genes are not formed by a periodic molecular process, such as a wave. Instead, each stripe is formed independently and the periodic pattern is "patched together" from individual stripes. The position of each stripe is determined by a different combination of gap-genes. In the hypothetical example of the two stripes of the pair-rule-gene "F" (Figure 30), the left "F" stripe is determined by the combination of the gap-genes "C" and "E," and the "F" stripe on the right by the combination of the gap-genes "D" and "E." Through interactions between the pair-rule-genes, the seven stripes of a given gene will be interlaced with those of another pair-rule-gene. In our example, "F" represses "G," hence "G" stripes can only be formed between the "F" stripes. In the *Drosophila* embryo, the pair-rule-gene fushi-tarazu is repressed by the even-skipped protein, and its seven stripes develop between the seven even-skipped stripes. The principles of the formation of a periodic pattern in the fly embryo are summarized in Figure 30.

Finally, controlled by the pair-rule-genes, 14 stripes of activity of the so-called segment-polarity-genes emerge. They are the last level in the hierarchy of segmentation genes. These 14 stripes already prefigure the pattern of the 14 segments of the larva and the adult fly. For instance, both even-skipped and fushi-tarazu activate the segment-polarity-gene engrailed. The engrailed stripes directly control the pattern within each segment. The line of cells transcribing the engrailed-gene will later form the posterior part of each segment (Figure 31, bottom).

Selector-Genes. The fine-tuning of the molecular prepatterns eventually leads to a division of the egg into segments. But at this stage the segments are more or less identical and the cells within the segments do not yet "know" which specific structures they are supposed to form. For example, are they to build the small teeth characteristic of the front segments, the broad bands of the back segments, or the structures of the head segments? Independent of the genes that create the spatial subdivisions are other genes that cause cells to adopt different states. These genes are called "selector-genes." The selector-genes encode transcription factors that activate groups of genes which exert their influence later in differentiation. One example of a selector-gene is the eyeless-gene, which determines the eyes; cells that produce the eyeless protein

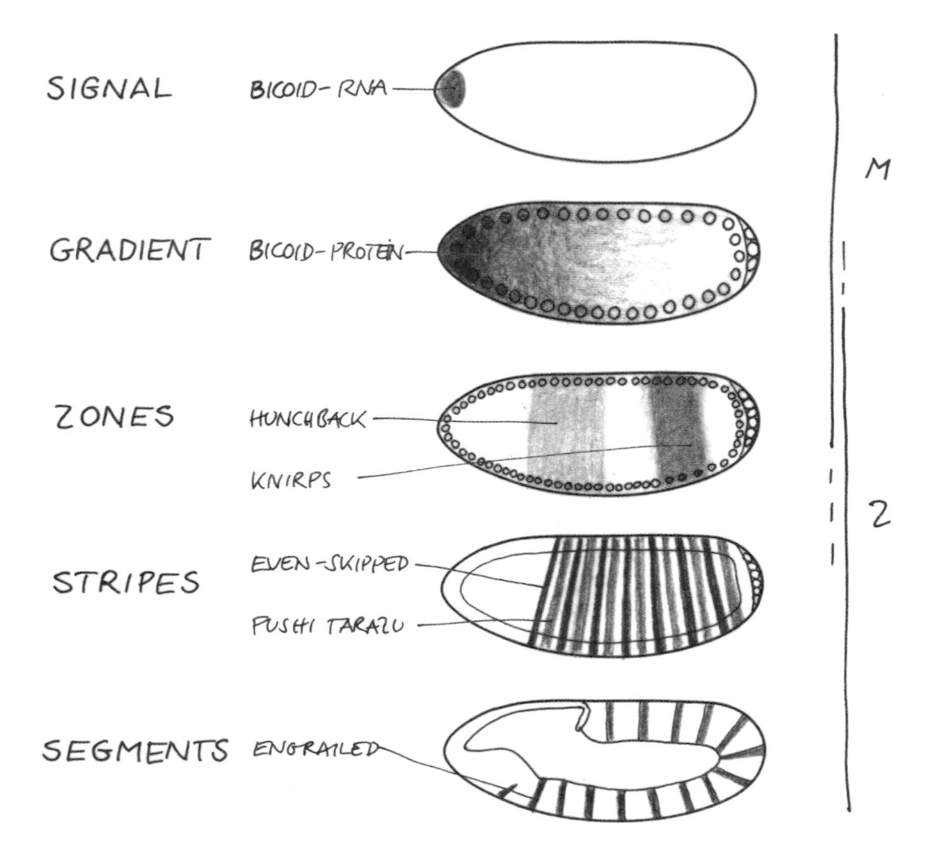

Figure 31. *Molecular Prepatterns.* A series of molecular prepatterns determines early *Drosophila* development. This figure shows the logic of hierarchical genetic regulation resulting in segmentation of the *Drosophila* embryo. A maternal signal (uppermost panel) is the localized mRNA (black) of the bicoid-gene. As this mRNA gets translated, a gradient of bicoid protein (red) emerges (second panel). Bicoid protein (in conjunction with a posterior gradient) controls the transcription of gap-genes in broad domains of the embryo. Here, hunchback (light gray) and knirps (dark gray) are shown as an example. The combinatorial activity of all the gap-genes determines the transcription pattern of the pair-rule-genes, for example, fushi tarazu (gray) and even-skipped (red). This is an important transition, as the broad contiguous gap-gene domains are transformed into a periodic pattern. The double segmental pattern of pair-rule-gene products is transformed into a segmental pattern of segment-polarity-genes, such as engrailed (lowest panel, red). Genes in this class are transcribed in a subset of cells in each segment. Thus, embryonic development of *Drosophila* involves the subdivision of the embryo based on gradients of maternal gene products (M), into increasingly more fine grained domains of zygotic (Z) genes that eventually prefigure the segmental organization of the animal.

will develop into eyes, even if cells in a wrong location are experimentally forced to produce the eyeless protein. Other examples of selector-genes are the twist-gene, which determines the mesoderm that will later form the musculature, or the tinman-gene, a member of a subgroup of musculature genes, which drives formation of heart muscle.

In the 1930s, American biologist Edward Lewis (1918–2004) discovered a group of genes that determines the characteristic features of the fly's body from front to back. These genes were called "homeotic genes." Mutations in these genes lead to the formation of structures that normally develop in different segments in the wrong position. For example, a homeotic mutation may result in the development of wings in segments that normally do not form wings, or in the formation of legs in the place of the antennae. Obviously, most of these mutations are lethal because the larval segments develop structures typical for another segment. Flies have eight such homeotic genes, which are organized into two gene clusters on the chromosomes known as the Antennapedia and Bithorax complexes, respectively. It is worth noting that the order of these genes on the chromosome corresponds to the order of the structures they influence from head to tail, such that the gene-specifying antenna will position before a gene specifying the wing, etc. Their products work in combination—only one gene is active in the front regions, then two, and then more and more genes become activated further toward the posterior of the embryo. They determine the character of the individual segments along the body axis of the animal.

The homeotic selector-genes code for transcription factors that contain a so-called homeodomain (see page 38). The region of the gene coding for the homeodomain is called "homeobox." Regions in which these so-called Hox-genes are repressed or activated are defined early in development by interaction of the segmentation genes, mainly the gap-genes. However, the activation of Hox-genes is stable while the gap-genes are only temporarily active. There are special mechanisms by which cells can remember the state of Hox-gene activity through many cell divisions.

Hox-genes have been found in modified versions in all animals studied so far, even in worms that are not segmented. In vertebrates, there are 13 different Hox-genes arranged in only one cluster, which occurs four times in different locations in the chromosomes. The discovery of the Hox-genes provided the first clear indication that animals of very diverse groups such as *Drosophila* and mouse are built according to a similar basic body plan. This similarity is based on homologous genes, which arose during evolution from a common ancestor.

3. Induction and Signal Transduction

As soon as the cell membranes are formed during *Drosophila* development, transcription factors can no longer spread from nucleus to nucleus by simple diffusion. At this point, the cells secrete proteins that can be recognized and bound by other proteins on the neighboring cells. Once a secreted factor is bound by such a receptor, the binding information is passed to the inside of the cell via a series of molecular switches until it is finally conveyed to the nucleus. Ultimately, the binding of a secreted factor to its receptor results in a change of the activity of one or more genes of the cell. Such signals can travel to adjacent cells, as well as over longer distances. Signal transduction is vital during the formation of cell patterns emerging from the interaction between adjacent cells.

During development, layers of cells often abut each other. One sends a signaling molecule that is received by the other, and in turn induces a change in the tissue. This process is called "induction." In other cases, signaling molecules spread within a layer of cells. Signaling molecules can also form concentration gradients in the extracellular space, and can act as morphogens influencing the cell's fate depending on their concentration (Figure 32). Often several signals reach the same cell, resulting in a competition or a response to a combination of different cues.

In the simplest case of signal transduction, the signaling molecule penetrates the cell membrane and directly binds to a transcription factor to regulate gene activity. This mechanism is typical for hormones and involves signaling molecules that can cross the membrane. But in most cases the signaling molecules are proteins that are too large to cross the membrane. In these situations, the process involves receptor proteins that bind the signal proteins as ligands at the cell membrane. The receptor proteins reach into the cell, and transmit the information of the ligand binding to the nucleus (Figure 33).

Several types of protein are involved in signal transduction. They exist in two states, either active or inactive. Activation generally entails a change in the three-dimensional structure of the protein. Binding of a ligand to a receptor activates that receptor. For example, activation can consist of two receptor molecules binding to each other. In many cases, activation results in a phosphate group being attached to the part of the receptor that reaches into the cell. Binding of another protein to this phosphorylated receptor will again activate this protein so that it can transfer a phosphate group as well. Proteins that transfer phosphate groups are called "kinases." Finally, the cascade of sequential activation reaches a transcription factor that represses or

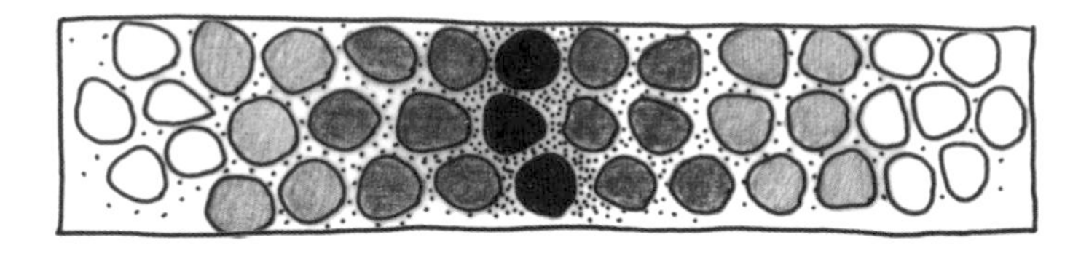

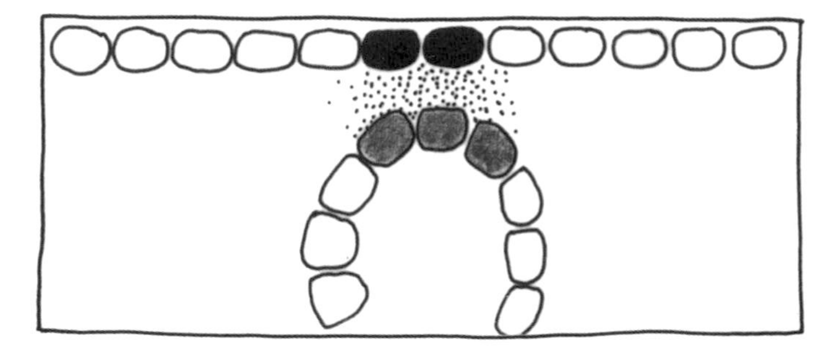

Figure 32. *Induction.* A morphogen gradient may result when a cell group (black) produces and secretes a morphogen. The morphogen diffuses in the extracellular space and influences neighboring cells depending on its concentration. Different thresholds create several zones (red and gray). The secreted morphogen may also reach and induce neighboring cell groups from the outside as shown at the bottom of the figure. This is how for instance the morphogen Dpp works in several types of tissue in flies and vertebrates.

activates one or more genes. Such signaling cascades need not always rely on the activation of proteins; in some instances, the goal is also reached by inhibition of a process that normally occurs spontaneously. Interestingly, there appears to be a limited set of types of signal transduction cascades that are used repeatedly within the same organism to induce tissue differentiation, form patterns, and initiate growth.

Examples of such cascades, named according to the gene encoding the signaling molecule, are the "Decapentaplegic," "Notch," "wingless," and "hedgehog" cascades. These signaling molecules are active early on during gastrulation in most animals and again later during processes such as the pattern formation of organs and tissues. The various signal transduction cascades differ with regard to the spatial range of the signaling molecule, as well as the manner in which that signal is passed on to the nucleus. The different chains are also employed in varying frequency and versatility within the organism.

Research on *Drosophila* has uncovered many components of such signal transduction pathways as gene groups with the same or similar phenotype.

Figure 33. *Signal Transduction* . The figure shows a simplified diagram of a cell receiving an extracellular signal. The signal (black), such as a morphogen, binds to receptor proteins (dark gray) located in the cell membrane and thus activates them (indicated in red). The proteins in turn activate a cascade of other proteins. The last step in this cascade is the activation of a transcription factor, which then activates the transcription of one or several genes in the nucleus. If one of the components of this signal transduction chain is defective or missing by a mutation in the respective gene, then the entire chain breaks down. This is why mutants of any components of these signal transduction cascades often have similar phenotypes.

One of the most remarkable discoveries in genetic research is that these signaling cascades are similar throughout the animal kingdom and may have quite different functions. One such example is the maternal dorsal gradient with Toll as the receptor and the dorsal protein as the transcription factor. This signal chain involves more than 10 genes. Their proteins are either involved in the production of the signal, or in passing it on to the dorsal protein, which enters the nucleus when activated. A very similar pathway using a Toll-like protein as receptor, and including many of the other 10 genes, is also operating in both flies and vertebrates in the activation of an innate immune response.

More often, however, similar signaling cascades take on similar functions in different organisms. Many genes encoding signaling proteins that were discovered in *Drosophila*, such as Decapentaplegic, wingless, hedgehog, and Delta, are also present in other animals. They play major roles in the pattern formation processes, although the gene names in *Drosophila* differ slightly from the names of the corresponding vertebrate genes. For example, in vertebrates wingless is called wnt, and hedgehog is called sonic hedgehog. Notably, some of these and other pathways were independently identified for their role in human cancers. This makes perfect sense, as many of these pathways control tissue growth, and thus tumors ensue when these pathways are misregulated.

Three examples of signaling pathways are discussed below.

Decapentaplegic (Dpp). Decapentaplegic is a Greek word meaning "15 mishaps." The name illustrates the fact that, depending on the severity, a mutation in the Dpp-gene may affect many different structures of the larva or the adult fly. The Dpp protein belongs to a class of important and widespread growth factors, known from cell biology and cancer research as transforming growth factor-β (TGF-β) or bone morphogenic protein (BMP). Early in *Drosophila* development, the Dpp protein emerges on the dorsal side and spreads toward the ventral side along the exterior of the cells (see Figure 29). On the ventral side, Dpp protein is inactivated by the extracellular short gastrulation protein, to result in a Dpp protein gradient that has its peak at the dorsal side. This gradient determines the dorso-ventral pattern of the larva. Once the mesoderm has folded in, the Dpp protein signals to the interior as well, and in turn, induces mesodermal cells adjacent to the dorsal epidermis to form the heart. Later, in the imaginal discs, Dpp acts again as an important morphogen. The various activities of Dpp as a morphogen and as an inductor are shown in Figure 32.

Delta–Notch. In this signaling pathway, a cell signals to all its neighbors. However, in this case the signal does not diffuse. The signaling protein Delta and the receptor Notch are both anchored in the cell membrane, therefore, only the adjacent cells can receive the signal (Figure 34). Initially, both cells have Delta and Notch protein at the surface. The activation of Notch by Delta in one cell results in a reduced production of Delta protein in those cells. This means that they will send fewer Delta signals back. In turn, more Delta protein is produced in the cell that receives fewer Delta signaling. This balancing feedback mechanism ensures that eventually only one cell produces Delta protein, whereas the neighboring cell represses it (Figure 34). Thus, Notch–Delta signaling makes initially similar cells different. During the formation of the nervous system, for instance, individual sensory cells arise in an initially homogenous cell sheet according to a certain pattern. Cells rich in Delta protein become sensory cells while those with a high amount of Notch form epidermal cells.

Hedgehog and Wingless. The hedgehog- and wingless-genes were discovered as segment-polarity-genes in *Drosophila*; their proteins function as ligands

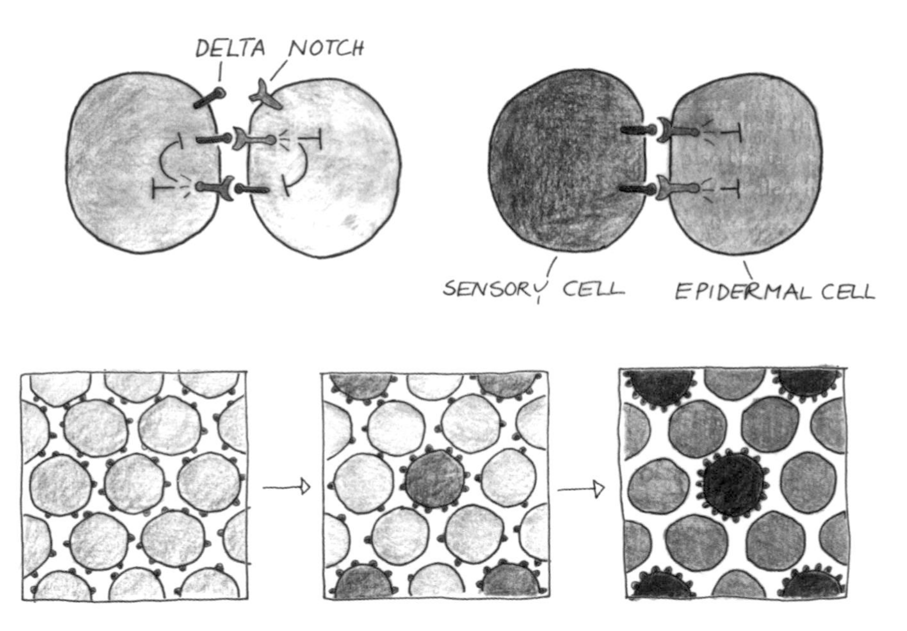

Figure 34. *Delta-Notch Signaling.* In this case, both receptor (Notch) and ligand (Delta) are anchored in the cell membrane. Initially, Notch and Delta are both present in the membranes of neighboring cells (left). Activation of Notch in a given cell results in this cell producing less of the Delta protein. This means that a neighboring cell will receive fewer Delta signal from that cell and consequently show a lower level of Notch receptor activation. As Notch normally represses Delta, this cell will make more Delta, resulting in increased Notch activation in the first cell, and so forth. This means that eventually neighboring cells will either produce Delta protein or show Notch activation. This is how "salt and pepper" patterns develop (bottom) and how initially similar cells can become different. They distinguish individual cells within a cell layer such as the epidermal cells and sensory cells. This figure has been redrawn from Bruce Alberts' (et al.) book *Molecular Biology of the Cell.* London: Garland Publishing; 4th edition, 2002.

and are secreted. In the early embryo, the synthesis of the hedgehog protein is activated in a stripe of cells by the engrailed protein. The hedgehog protein then diffuses and is bound by receptors on the neighboring row of cells. These cells activate the gene encoding the transcription factor Ci, which in turn triggers the synthesis of the wingless protein. Wingless protein then signals back to activate engrailed protein production. These interactions create a positive feedback mechanism that ensures the synthesis of the Ci and engrailed proteins in two neighboring rows of cells. This pattern is stable and remains intact until the adult form is taking full shape (Figure 35).

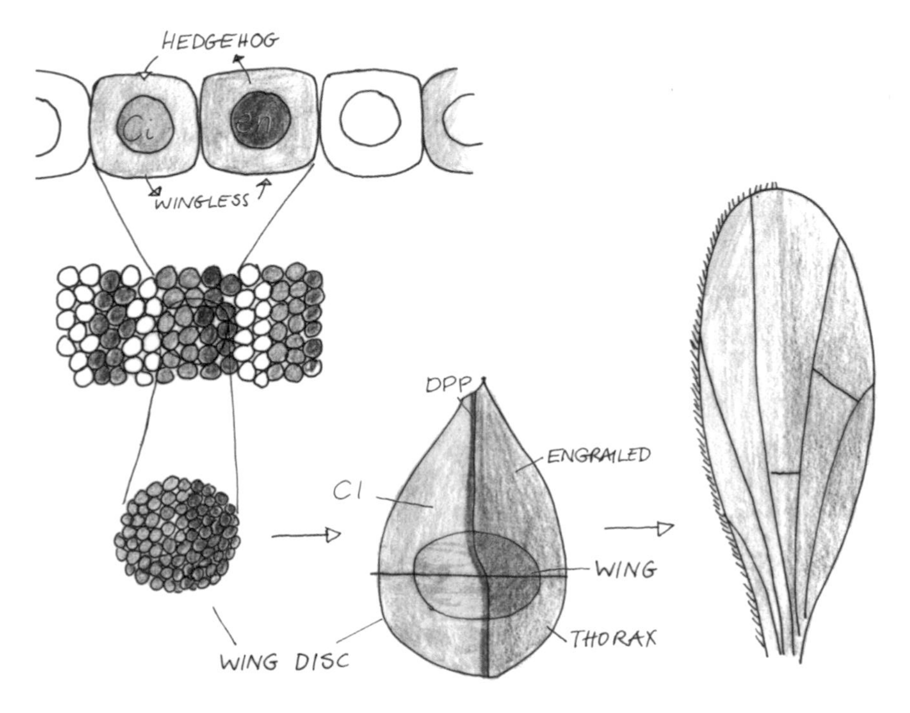

Figure 35. *Wingless-Hedgehog Interaction.* During segmentation (uppermost panel), neighboring cells use the secreted wingless and hedgehog proteins to distinguish themselves. In this case, they enhance each other. Thus, after the original distribution has been determined by pair-rule-genes, wingless activates hedgehog transcription in neighboring cells, from which hedgehog protein diffuses and causes more wingless production in the cell it originated from. Therefore, they form stable neighboring relationships which are maintained by the transcription factors Ci (gray) and engrailed (red). The imaginal discs (middle lower panel) develop from this double row of cells and stably maintain the relationships. The wing disc shows a prepattern of Ci (gray) and engrailed (red), which eventually develops into the front and back portion of the wing, respectively.

As mentioned before, the legs and wings of the fly develop from cell groups that are set aside early on within the segment precursors (see Figure 21). These cells develop into the imaginal discs that are destined to form the structures of the adult fly. The imaginal discs emerge from such engrailed and Ci protein expressing rows (Figure 35). The cells continue to divide and form the imaginal discs while the neighboring epidermal cells do not divide any further but just grow in size (see Figure 21). As inherited from their origin, the cells in the front part of the imaginal discs will produce the proteins of the Ci and wingless-genes, while the cells in the back part will continue to produce the engrailed and hedgehog proteins (Figure 35).

Morphogenetic gradients, combinations of factors, and signal transduction cascades lead to a subdivision of the embryo into increasingly smaller units. The action of selector-genes finally defines cell groups and determines their future fate. The general principles of these developmental processes are now quite well understood, although there are of course many details that need further clarification. This is particularly true for processes at later stages of development, which are harder to study because many of the genes have already been active earlier, and thus, their mutants do not survive to the stage of interest. However, for most cases, it is still poorly understood what it means exactly that cells are destined to a particular fate. What outcome do selector-genes evoke? Which genes are activated when the cells begin to differentiate? How do these genes trigger shape changes that eventually lead to the finite form, and to subtle differences such as the distinction between the front and back of the wing in *Drosophila*? The processes leading to cell differentiation can only be understood once we know what determines the cell's final shapes in the first place.

CHAPTER VI

Form and Form Changes

DURING DEVELOPMENT, ANY CHANGE IN CELL SHAPE IS PRECEDED BY A change in gene activity. It is the cell's origin and environment that determine which transcription factors are active within a cell and, hence, which genes are turned on, and which proteins are produced.

All embryonic cells have a history—they emerge from the division of a mother cell—and they all are in contact with neighboring cells. Responding to molecular prepatterns, cells may change their shape; may leave or join a cell group; they may grow, divide, or eventually die. Each individual cell in a given tissue is constantly influenced by many extracellular signals. These signals may counterbalance each other to help stabilize the cell's state. On the other hand, it is possible that only a minor fluctuation in a single signal can change the state of the cell.

Cell shape changes and cell movements during gastrulation are crucial for the formation of the embryo. Starting as a homogenous layer of cells or a cell ball, the embryo becomes multilayered and elongated. Interestingly, the repertoire of cellular processes and the molecular mechanisms is limited. Consequently, the same mechanisms that create embryonic form during gastrulation are also used subsequently for tissue and organ formation. Furthermore, many molecular pathways were conserved to a large extent during evolution; as a result, these pathways can be found in organisms as simple as fungi and as complex as vertebrate animals.

1. Cells and Tissues

In animals and most plants, cells are organized into tissues. Tissues can be highly structured, such as the two-dimensional sheets of cells known as epithelia, which line the interior of the gut or the skin; alternatively, cells can be loosely scattered in a type of tissue, which is referred to as a mesenchyme. In general, tissues are embedded in a gel-like or fibrous material known as the extracellular matrix (Figure 36).

The shape of a cell is determined by its inner skeleton—the cytoskeleton, by its connection to neighboring cells, and by the extracellular matrix it attaches to. The establishment and maintenance of cell shape involves a great number of specific, form-giving proteins. Many of these proteins are polymers composed of one or a few subunits that are arranged in long chains, which often form bundles. Important for their function is the way they connect to each other or to other structural proteins.

Cytoskeleton. The cytoskeleton comprises a group of several types of flexible protein fibers that extend through the cell. The so-called microfilaments consist of long chains of actin molecules which are organized into bundles and tightly knit three-dimensional networks of fine threads. They are particularly dense directly underneath the cell membranes, forming a microfilament "cortex," which stiffens and strengthens the cell membranes (Figure 37). Actin chains can shorten or elongate by addition or removal of actin molecules at either end of a thread. The actin chains associate with proteins called myosins. Myosins are the so-called motor proteins that are able to move the actin chains. These very same proteins are used in highly ordered fibers to generate the force in muscle cells and operate in a similar way within the cytoskeleton to divide a cell or to change its shape. When a cell is

Figure 36. *Tissues.* Epithelia (top) are densely packed cell layers with an inside and an outside. The epithelial cells sit on top of an extracellular basal membrane. Examples of organs formed from epithelia are the skin or the gut. Mesenchymes (bottom) are loose tissues that are embedded into the extracellular matrix, such as the cells giving rise to bone and cartilage. Many tissues can be regarded as an intermediate between the extreme forms illustrated here.

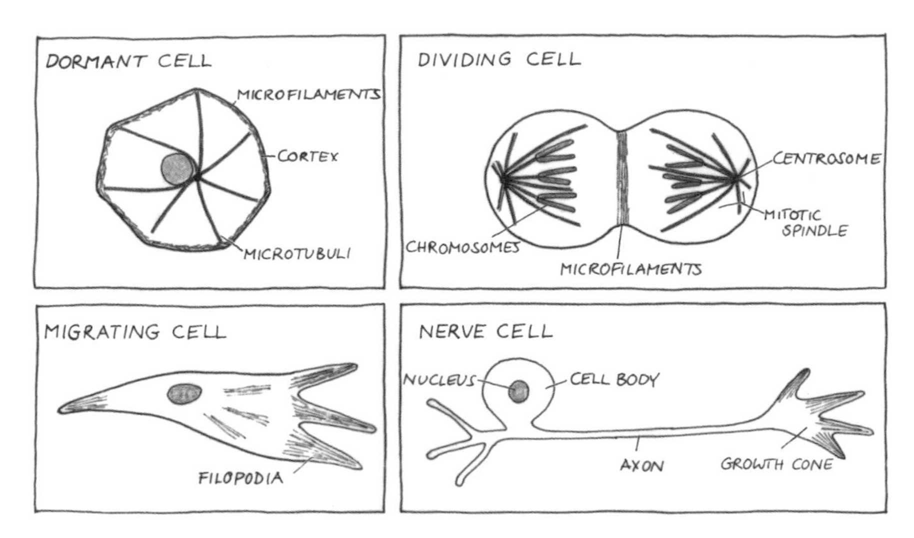

Figure 37. *The Cytoskeleton.* Microtubules (red) originate from the centrosome, their function is to either stiffen the cells (top left) or to transport chromosomes to the daughter nuclei as spindle fibers during mitosis (top right). They also serve as tracks for transport in the axons of nerve cells (bottom right). Microfilaments (gray) are bundles of actin fibers that create a denser mesh at the inside of the cell membrane, known as cortex. After mitosis, they divide the daughter cells by forming a contractile ring (top right). They also form filopodia that enable motility and shape changes of migrating cells or the growth cone of the nerve extensions (bottom panels).

moving, it forms extensions called filopodia in which the microfilaments change continuously (Figure 37).

Microtubules are stiff, helical chains of tubulin molecules that form a hollow tube. They usually originate from an organizing center, the centrosome, and elongate or shorten by either adding or removing tubulin molecules at the free ends of the tube. Special proteins can attach the free ends of the microtubules to the actin cortex. The result is a network of fibers that shapes and stabilizes the cell.

The microtubules have two additional important functions. First, during cell division, they pull apart the duplicated chromosomes. To this end, the microtubules form a spindle that usually originates from two centrosomes. The microtubules of the spindle attach to special points on the chromosomes (Figure 37). After each chromosome pair is attached to the microtubules from the opposite spindle pole, they are pulled toward each centrosome; this ensures that each daughter cell receives one complete set of chromosomes.

Second, the microtubules act as transport tracks for the distribution and localization of many proteins and organelles. Motor proteins can transport these components either to or from the center of the cell. This function of the microtubule cytoskeleton plays a major role when cells become polarized; in other words, when one part of a given cell becomes structurally different. For example, in the *Drosophila* egg, the bicoid RNA is transported to the front end of the egg cell along microtubule tracks.

In addition to microfilaments and microtubules, there are also other fibrous networks that fulfill special functions within certain cell types. The bundling, lengthening, networking, and stabilizing of the cytoskeletal fibers are regulated and modified by numerous proteins. The versatility of the cytoskeleton allows for the formation of an extraordinary variety of cell forms.

Cell Adhesion. Cells bind to one another via membrane proteins that build connections between cells. Cadherins are one type of such membrane protein. Large parts of the cadherin molecule extend into the extracellular space and bind to cadherin proteins on the neighboring cell. The part of the protein that reaches inside the cell anchors the cadherin molecules to the actin cytoskeleton of the cell cortex. In this manner, the cells of a tissue are held firmly together. There are several different types of cadherin that characterize specific cell types. Also, cadherins usually bind to cadherin proteins of the same type. In addition to the cadherins, there are numerous other cell adhesion proteins.

The mechanism of adhesion of similar cadherin types helps to ensure that similar cells remain attached to each other, thus stabilizing the tissue. Cells with a different type or density of adhesion molecules become segregated from the other cells.

Extracellular Matrix. The extracellular matrix is the substance between cells and consists of several different types of proteins. The most common of all extracellular proteins of animals, collagen, is a component of the extracellular matrix of cartilage and connective tissue. Many other proteins of the extracellular matrix are the so-called glycoproteins, which contain sugar components. These sugars are responsible for the proteins' gel-like properties. Other proteins, such as fibronectin and laminin, have rather complicated structures consisting of several subunits that interact in complex combinations. These proteins are present in the basal membrane of epithelia. They are important for the movement of cells in the extracellular matrix. Membrane proteins of the cells, called integrins, bind to fibronectin, thus anchoring the cells to the basal membrane.

2. *Form-Building Movements*

Important form-building processes involve the invagination of epithelia, thus resulting in the formation of multiple layers of cells. Epithelial tissues can also transform into loosely connected cell complexes called mesenchymes. In other cases, the mesenchymes can also change their shape to convert into epithelial tissue. Eventually, cell movements may lead to shifts of cell groups and new arrangements within the tissues. Many organs are built from epithelia, sometimes multilayered, such as the intestines, the skin, and the blood vessels. Others are formed from mesenchyme such as the muscles and the blood. Movements of an entire group of cells are the result of cell shape changes driven by the dynamic modifications of the architecture of the cytoskeleton. In addition, induction processes can also lead to a change in the repertoire of adhesion proteins at the surface of a cell, as well as locally trigger cell division and growth.

Invagination. In the *Drosophila* embryo, cells arrange themselves in multiple layers during the process of gastrulation when a broad band of cells folds in to eventually become the mesoderm. This folding-in process is referred to as invagination. The cells that undergo invagination develop microfilament rings in the part that faces the outside of the embryo. These rings contract and constrict the cells on one end, and thus bend the cell layer. This leads to the invagination of the cell layer into the embryo (Figure 38). The cells that are folded-in experience a different environment from those still remaining on the exterior. In the interior, the cell complex loosens and spreads along the inside of the ectoderm. Invagination also plays a major role during the formation of organs such as the neural tube of vertebrates, glands, and sensory organs (Figure 43). The hindgut of the *Drosophila* embryo is formed by the invagination of a cell plate at the posterior pole of the embryo, which later forms into a tube.

Ingression. Ingression of individual cells or entire groups of cells is another way for cells to get inside the embryo. Individual cells or groups of cells separate themselves from the epithelial layer and move into the interior of the embryo where they can associate loosely with other cells (Figure 38). The nervous system of the *Drosophila* embryo develops in this manner. As described previously (see Figure 34), the neuroblasts, which are the precursors of the nerve cells, are defined by Delta–Notch signaling. The neuroblasts shift toward the inside of the embryo and, after several cell divisions, form

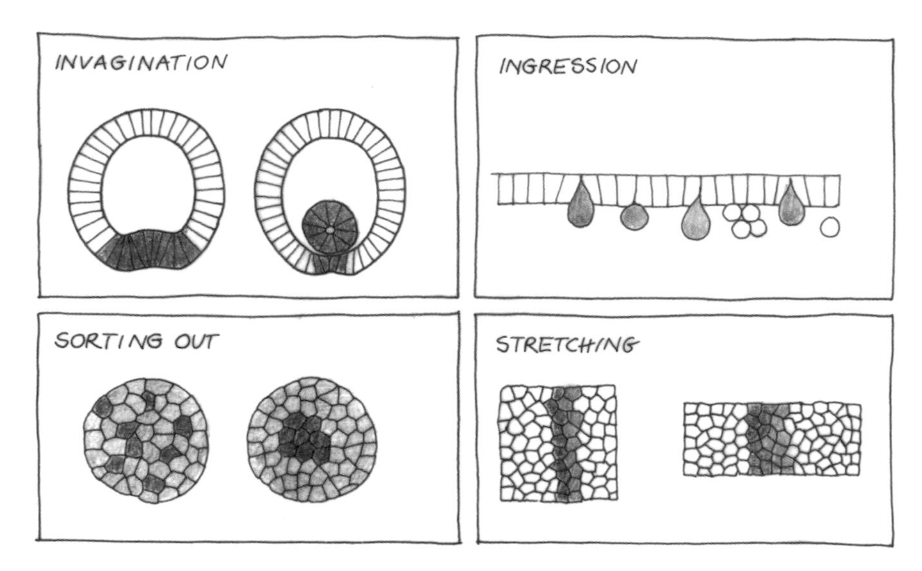

Figure 38. *Form Changes in Tissues.* Invagination (top left) is a result of contracting microfilaments and takes place during the development of the *Drosophila* mesoderm, as shown in this example, or during neurulation of vertebrates (see Figure 43). By ingression (top right), individual cells move to the inside of an embryo, such as the neuroblasts in the *Drosophila* embryo or in the mesoderm of the chick (see Figure 46). The sorting out of cells (bottom left) is based on varying degrees of cell adhesion. Cells that are more sticky will end up in the center of this exemplary tissue. Extension or stretching (bottom right) of a cell layer while keeping intact the neighboring relations occurs as cells slide past each other. Examples of this stretching process include the gastrulation of the frog or the extension of the germ band of the *Drosophila* embryo.

the nerve cells, the ventral nerve chord, and cells surrounding the nervous system. Precursors of the front part of the central gut develop via the ingression of blastoderm epithelial cells. Later, from an initially loose collection of cells, they form the epithelial tubes of the digestive tract. The multiple layers of the chick and mammalian embryo are also created by ingression of cells (see Figure 46).

During gastrulation, extensive rearrangements take place, resulting in a change in the relative location of the cells with regard to one another. In *Drosophila*, the multilayered germband forms in this way. The geometrical shape of the trunk region of the embryo changes from a compact shape shortly after gastrulation to a more elongated streak, called the extended germband (Figures 20 and 38). Only during this extended state, segment borders develop and cells forming the central nervous system move inside.

Again, these rearrangements are the result of cell shape changes caused by microfilaments.

Cell Migration. The directed migration of cells and extending nerve fibers follow signals that either attract or reject them. These signals are sensed by receptor proteins within the membrane of the migrating cells. In these cases of signal transduction, the cytoskeleton is influenced directly by the signals without them first affecting the cell's transcriptional activity. By means of such controlled cell shape change, nerves move along complicated tracks that will eventually bring them from the central nervous system to the organs or the musculature. Likewise, sensory nerve cells originating in the sense organs will extend their fibers from the sense organ to the brain.

When nerves grow, the body of the nerve cell stays at its original location while it sends out long extensions called axons. The axon is stiffened by microtubules. The tip of the growing axon, the growth cone, extends filopodia that move and explore the environment, searching for guidance signals (Figure 37).

3. Cell Division, Growth, and Death

The development of all multicellular organisms is accompanied by cell divisions that occur in an ordered and controlled manner. During the embryonic development of most animals, cell division is initially not coupled to cell growth. Instead, development begins with a subdivision of the egg called cleavage. True growth itself cannot take place until the larval stage when the organism first begins to receive its nutritional supply from the outside. Early divisions are coupled to cell growth only in embryos with very large yolk supplies such as birds and reptiles. After implantation in the uterus, mammalian embryos continuously receive nutrients through the maternal blood and cells grow before each division.

For most animals, the early stages of development are characterized by a rapid succession of cleavage divisions. Later in development, cell divisions are initiated by outside signals known as growth factors. With rare exceptions, cell divisions take place mainly in cells that are not yet fully differentiated. Before differentiation, cell divisions are stopped and this stop in general is irreversible.

Cell Division. Cell division begins with DNA replication whereby two copies emerge from each chromosome. Mitosis, the distribution of the copied chromosomes or chromatids to the two newly forming daughter cells, is initiated only after this process has been completed. The chromosomes condense and the spindle pulls the two chromatids in opposite directions. Between the separated chromatids an indentation is formed by a microfilament ring that contracts and thus separates the two daughter cells (Figure 37).

The exact order of steps during cell division is carefully controlled. First, the replication of the DNA needs to be initiated. This involves certain proteins known as cyclins, which bind to kinase proteins and activate them. This causes the activation of DNA replication. Once replication is initiated, the cyclins disappear and the kinases become inactive. A specific cyclin–kinase complex that involves newly synthesized cyclins controls the transition to mitosis as well, ensuring that mitosis does not start before the chromosomes have been completely replicated. If cell division is coupled with growth, the volume of the cell increases continuously and the control mechanisms make sure that the division takes place only when growth has been completed.

The process of cell division can be modified by skipping one or two steps. For example, after the early divisions of the nuclei of the *Drosophila* embryo, no membranes are formed between the cleavage nuclei. This creates a multinucleated cell, a syncytium (see page 46). During larval development, the larval cells grow and their DNA replicates, yet the usual mitosis and the corresponding cell division do not take place. Still, the cells keep growing larger and larger, compiling much more DNA than the normal diploid genome and become polyploid. This may lead to a formation of the giant chromosomes described in Chapter III.

Meiosis. When the germ cells, the gametes, develop, the number of their chromosomes is reduced by half so that the gametes have only one copy of each chromosome, instead of the usual two. During fertilization, the haploid egg and sperm combine to create a diploid zygote, the beginning of life for a new organism. The reduction of chromosome number is achieved in a process called meiosis. It is the result of two successive cell divisions. Before the first division, the chromosomes replicate just as in mitosis to form two chromatids that stay connected at the sites where the spindle fibers attach. But unlike during mitosis, where the chromatids are pulled apart by the spindle, the homologous chromosomes arrange in pairs. At this stage, crossing over (see Figure 12) between the chromatids of the homologous chromosomes takes place and segments of chromatids are exchanged. This results in a new combination of genes from the two initial chromosomes. Every pair of chromosomes experiences two to three of these crossing over events. After recombination, every one of the four chromatids has a different genotype and is thus unique. The separating spindle then distributes the chromosomes to the daughter cells in a way that each receives only one of the homologous chromosomes. After a second, mitotic division, four haploid germ cells emerge (Figure 39), each with one set of chromosomes. In males, all four give rise to sperm. In the female germ line only one of these four products of meiosis develops into an egg cell, while the other three are pinched off as polar bodies and eventually die (Figure 51).

Cell Death. When the divisions of cells in a tissue are completed, a stop signal tells the cell to rest. The stop signal has to be inactivated by a growth factor if that cell is meant to divide again at a later time. If both stop and growth factors are missing, then the cell dies. This form of programmed cell death, known as apoptosis, erases all cell content and leaves no traces behind. Apoptosis is not an accident, but in fact can be induced by certain factors as part of the regular repertoire of pattern formation. For example, some cell

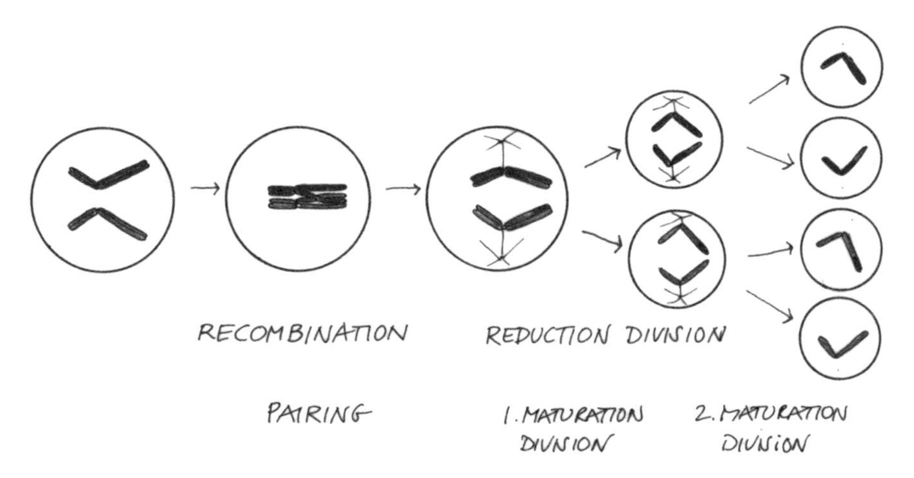

Figure 39. *Meiosis.* During meiosis, the number of chromosomes is halved in two consecutive cell divisions so that the gametes only receive one copy of each chromosome pair. At first, the chromosomes are doubled, just like at the beginning of mitosis. At this time they are arranged in pairs and a crossing over between homologous chromosome segments takes place, resulting in recombination of parts of the chromosomes. The following division separates the homologous chromosomes. The daughter cells divide once again in a "normal" mitotic division. The four cells resulting from this division are haploid and, due to recombination, are distinct from each other. This figure only shows one pair of homologous chromosomes.

divisions in the worm *Caenorhabditis elegans* create new cells that shortly thereafter disappear through apoptosis. Apoptosis is also important during the development of the nervous system in many animals because initially many more nerve cells than required are produced. The surplus is eliminated by apoptosis. Finally, apoptosis is triggered by certain signals secreted by unhealthy cells so that they are eliminated.

Growth. In growing tissues, the volume of the individual cells increases continuously until the cell divides. New molecules are built into or bound to existing structures such as membranes and organelles. The organelles grow, become bigger, and finally divide. The formation of new cell structures occurs to a large extent by increase in size or replication of what already exists. Every cell emerges by division of an existing cell, so that in theory, all cells can be traced back without interruptions to one single original cell.

The form-building mechanisms in the embryo require that signaling molecules only have to travel short distances between cells in order to exert their influence. The size ranges of form-building processes are limited by molecular parameters such as protein stability or diffusion constants of morphogens. It is fascinating that also in very large animals, the essential body organization takes shape in an embryo, which is only a few millimeters large. All further development is mainly growth in cell size or cell divisions, although changes of the proportions of body parts can still be quite drastic. Before final differentiation, indeed many shape changes and elaborations within individual raw structures still occur as they form the final shape of the animal.

In *Drosophila*, morphogenesis and growth are separated to an extreme degree. The larva that is shaped during embryogenesis is only 1 millimeter long and grows 20 times larger, without any major shape changes, solely by an increase in cell size. During this period of growth only the cells of the imaginal discs divide regularly, but morphogenesis does not occur before the imaginal cells differentiate during the pupa stage (see also Figure 21).

Stem Cells. In the adult animal, growth is ceased, and only cells in special tissues are constantly renewed. These cells are produced by stem cells. Stem cells are undifferentiated somatic cells that can both replicate themselves and create differentiated cells (Figure 40). Some stem cells can only generate one particular cell type. Other so-called multipotent stem cells can produce several types of cells. Tissue such as the skin, the intestine, and the blood are constantly being renewed from stem cells. Stem cells can also be found where there is less renewal such as in muscles and the nervous system. Although

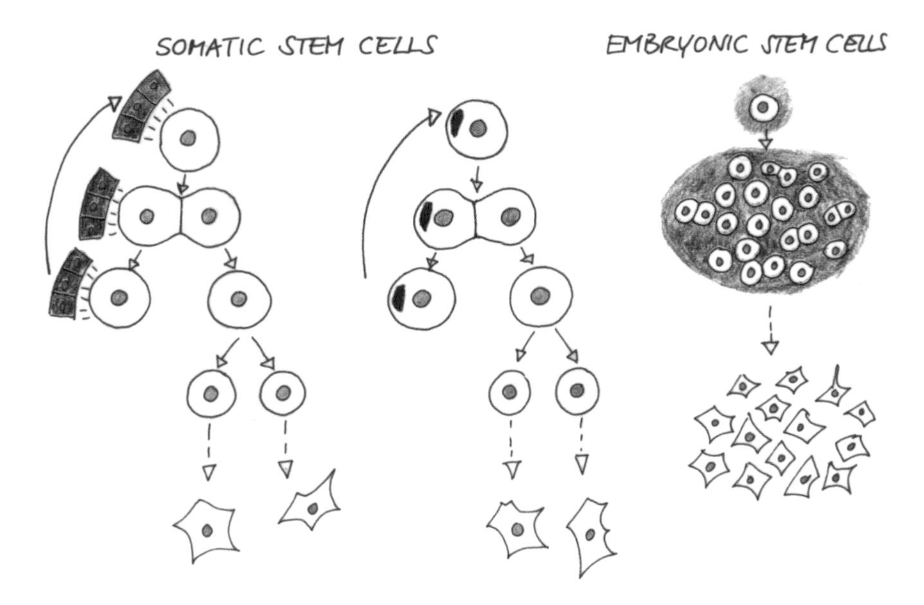

Figure 40. *Stem Cells.* The divisions of somatic stem cells (left and center) result in a stem cell as well as a cell that will give rise to a specialized cell type, such as skin or nerve cells. This cell can divide several times before it finally differentiates. The asymmetry of the stem cell divisions is either due to outside signals working on one side of the stem cell only (left) or due to localized components that are passed on to the stem cell, but not to the daughter cell (center). In media containing certain growth factors, embryonic stem cells (right) multiply while remaining stem cells (see Chapter VII). They differentiate if the growth factors are removed and then develop into a mixture of various cell types.

stem cells are undifferentiated, their ability to form different cell types is already limited. They seem to behave like the cells in the primordia of these organs, before these grew and differentiated.

Somatic stem cells show a special form of unequal division that produces non-identical daughter cells. This happens, for instance, when components, such as a localized RNA, in the original cell are distributed unevenly between its two daughters. In other instances, the dividing stem cells receive signals from neighboring cells that prevent the differentiation of one of the daughter cells. The cell in contact with these neighboring cells, which are known as the niche, remains a stem cell, while the other daughter cell facing away from the niche will give rise to differentiated cells (Figure 40). Usually, the offspring of the stem cells divide again several times before the final differentiation into specialized cell types occurs. A special kind of stem cells is the germ-line stem cells, from which eggs and sperm are made. These stem cells are maintained while unequal divisions create the egg and sperm precursor cells. In a sense, germ cells can be seen as the ultimate stem cells, because from them a whole new organism is made. Special components known as the germ plasm often play a major role in maintaining the germ line. Another special kind of stem cells, the embryonic stem cells or ES cells, originates from the early mammalian embryo. They multiply as pluripotent stem cells (Figure 40) and can give rise to any cell of the organism (see Chapters VII and VIII).

CHAPTER VII

Vertebrates

OF OVER 1 MILLION SPECIES OF ANIMALS, ONLY A FEW ARE SUBJECT TO intensive research in developmental biology. This is not because we want to know exactly how a fly or a frog develops, but rather we study flies and frogs as examples for the development of animals in general. Such animals are often called model organisms. Of course, the hope is that results from experiments on model organisms will be of general importance also to human development—although for many researchers this point does not play a major role. Some model organisms, such as the sea urchin, the fly *Drosophila*, and the worm *Caenorhabditis elegans*, are invertebrates and are therefore only distantly related to humans. Traditionally, vertebrate model organisms, such as the frog *Xenopus laevis* and the chicken play a major role for biological research. Their eggs are easy to obtain, and, more importantly, their embryos develop outside of the mother's body. The size of both the chicken and frog embryos allows crafty experiments that involve the isolation, transplantation, and combination of embryonic parts. In the 1920s, such experiments on newt embryos led to the discovery of the embryonic organizer, which influences the development of surrounding tissues (see Chapter II).

However, in both frogs and chicken, it is not easily possible to obtain mutants in which just one gene is turned off. In this respect, the zebrafish *Danio rerio* is more useful, because it is almost as well suited for genetic research as the fly *Drosophila* and the worm *Caenorhabditis*. In addition, its eggs are transparent and, therefore, the developmental processes can be directly observed in the living embryo. Conducting similar studies with mammals, like the mouse, is rather complicated because mouse embryos develop inside the mother's body and cannot be observed directly. Also, breeding mice results in only few offspring in a litter. However, unlike embryonic cells from the chicken, frog, and fish, cells from early mouse embryos can be isolated, cultured, and modified genetically. When transplanted back into an embryo, these embryonic stem cells can give rise to any part of the body of the resulting mouse. Embryonic stem cell cultures make it possible to turn off known genes in a directed manner. This process is referred to as knocking out genes and offers unique opportunities for genetic analysis. In addition, since the mouse is a mammal, like humans, knockout technologies are of particular importance for modern medical research.

1. Frogs, Fish, and Birds

Although adult vertebrates look very different from each other, they show many similarities in general body organization (Figure 41). For example, a

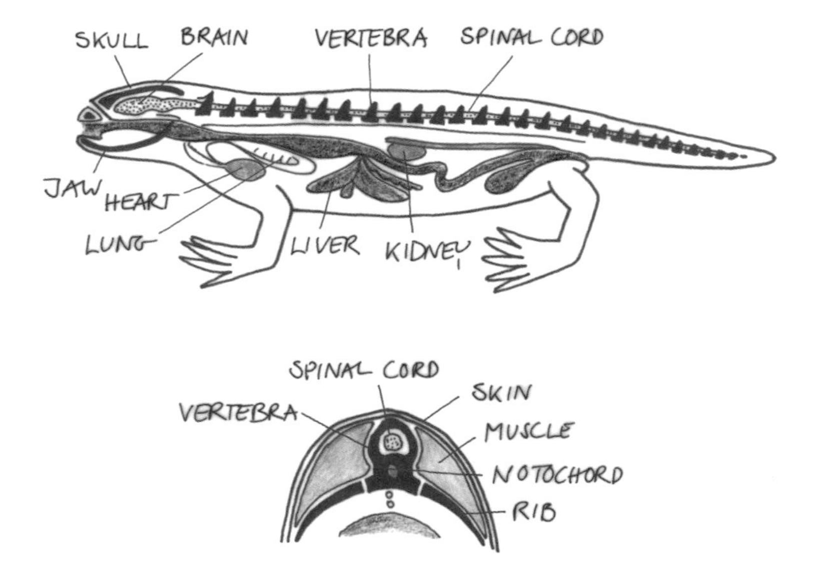

Figure 41. *Body Organization of a Vertebrate.* The name of this animal group stems from the vertebral column that protects the spinal cord and stabilizes the body. It still contains remnants of the notochord, an embryonic structure providing support of the body axis. The notochord gave the name to the chordates, which include vertebrates, jawless fish, and ascidians. Additional vertebrate features are the skull and the jaw, which develop from cells of the neural crest. The bottom of this figure shows a transverse section through the trunk region of a prototypical vertebrate. The organs of the digestive tract as descendents of the ectoderm are shown in red. The mesodermal organs are marked in gray. The limb bones are not illustrated here. This figure has been redrawn from Alfred Kühn's (et al.) *Allgemeine Zoologie*. Heidelberg: Springer, 13th edition, 1959.

rod-like structure called the notochord stiffens the body axis in all vertebrate embryos. It underlies and supports the neural tube that runs along the anterior–posterior body axis. The neural tube is the forerunner of the spinal cord and is broadened in the head region to form the brain. The nervous system is protected by the vertebral column in the trunk and the bones of the skull in the head. The muscles of the trunk develop from segmented blocks of tissue, called somites, which are located on the right and left sides of the notochord. The digestive tract and heart develop on the ventral side of the notochord. The heart pumps blood through the body in a closed system of blood vessels. These common features of the vertebrate body organization are particularly obvious in the embryos at a stage right after the formation of the major organ precursors (Figure 42). The limbs, feathers, scales, hair, skull, horns, claws, and other features that distinguish one vertebrate group from another develop only later.

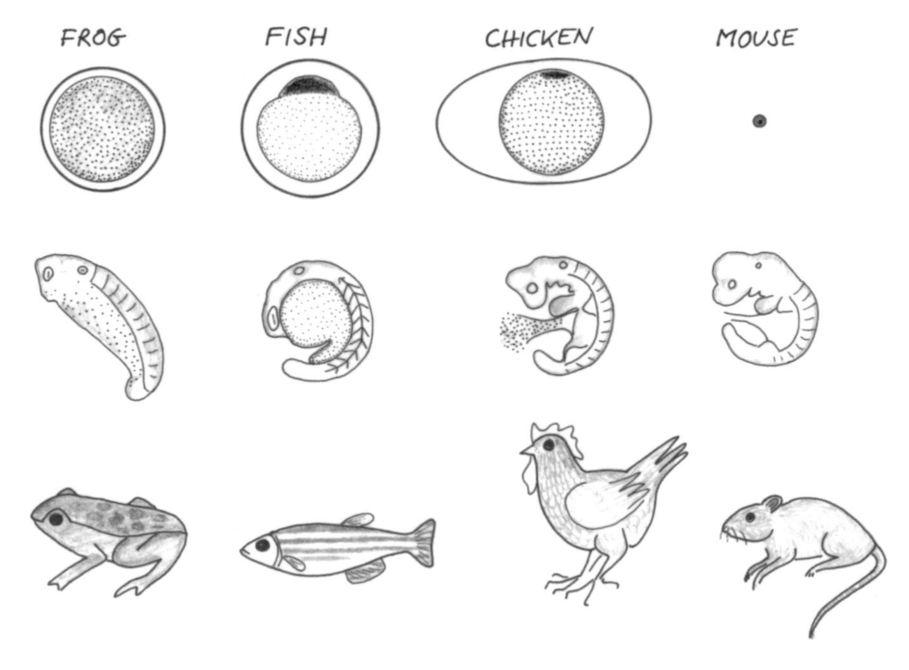

Figure 42. *Eggs, Embryos, and Adult Forms of Various Vertebrates.* The similarities among embryos are striking (middle row), yet there are great differences among the eggs (top row) and the adult animals. Eggs of frogs, fish, and chickens are very large and contain a lot of yolk as nutrient (stippled), while mammals, such as mice, have tiny yolkless eggs because the maternal organism feeds the embryo. When the young of egg-laying animals hatch they are about as big as the egg itself.

Profound differences are apparent in the early embryonic stages of different vertebrate classes. These differences often reflect different reproductive strategies; for example, the extent to which the parents care for the young or not, or whether a vertebrate class has adapted for developing on land by protecting the egg from drying out. Birds have huge eggs consisting mostly of yolk for the nutritional supply of the embryo. The embryo develops from a tiny disc sitting on top of this yolk. Frogs and fish produce many more eggs than birds, although they are much smaller in comparison. Their eggs are also rich in yolk. In fish eggs, as in bird eggs, the embryonic cells sit on top of a large ball of yolk, whereas in frog eggs the yolk is distributed within the embryonic cells. In the case of mammals, the mother directly supports the development of the offspring until birth. Mammalian embryos begin their development in a tiny yolkless egg. An important feature of this early stage in mammalian embryos is the differentiation of special cells that later give rise to parts of the placenta. The placenta will nourish the embryo through the mother's organism. The actual embryo develops after implantation in the mother's uterine wall.

As in other animals, the development of the vertebrate embryo begins with cleavage divisions that create a blastula consisting of many undifferentiated cells. During gastrulation, these cells are allocated to the three germ layers: ectoderm, endoderm, and mesoderm. In the case of birds and mammals, not only do embryonic structures develop, but also the tissue and coverings that protect and feed the growing embryo. These extraembryonic tissues disappear later, after they have served their purposes. Thus, the differences in embryonic organization are to a large extent due to the way the nutrients reach the embryo. The processes are very complicated, and it takes much experience and patience to recognize and classify the emerging structures of the various embryos.

Homologies. A comparison of the development of different vertebrates is interesting and helpful for a better understanding of the processes. Although quite different in their final appearance, many similarities between the structures of different species can be inferred based on homology, that is through their common evolutionary origin. For instance, the paired fins of fish, the paired wings of birds, and our paired arms and hands are all homologous structures. An important discovery was that the genes involved in the development of different species show even greater similarities than the embryonic structures. Therefore, the distribution of the products of such homologous genes can serve as a convenient marker for recognizing homologous embryonic structures. Using such markers makes it easier to compare embryos of different vertebrates. In addition, a gene discovered, for example, in the frog often has a similar function in the development of the chicken or the mouse. Therefore, knowledge gathered from one organism can readily be transferred to another.

The Development of the Frog Xenopus laevis. As previously mentioned, frog development is difficult to observe directly because the egg is not transparent, yet it has been studied intensively. The egg of a frog has a clear top and bottom that later roughly correspond to the front and rear of the embryo. During its early development, the frog egg divides by synchronized cleavages. The first cleavage division separates the egg into right and left halves, the second into a dorsal and ventral part, and the third division is located at a right angle to the top-to-bottom axis. Later divisions are not as clearly defined. The blastula then forms, consisting of several layers of cells that surround a hollow center. All cells contain yolk, yet those on the bottom side, which will later form the intestinal tract, contain distinctively more (Figure 43).

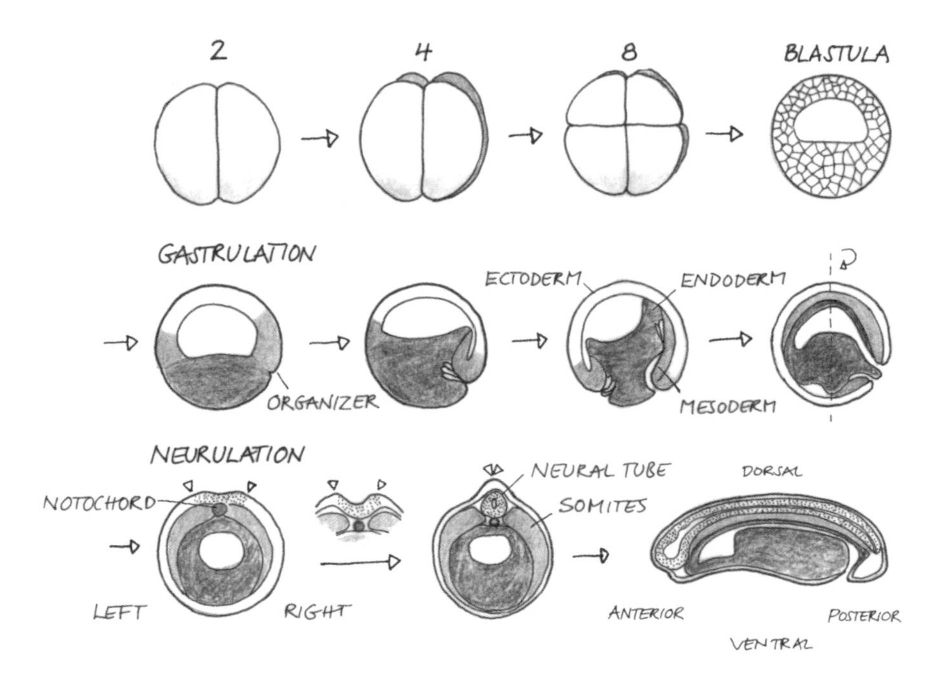

Figure 43. *Development of the Frog.* Cleavage (top row) subdivides the egg into increasingly smaller cells, giving rise to a fluid filled blastula. Gastrulation (middle row) begins on the future dorsal side at the so-called organizer. Along a ring-shaped zone, endodermal cells (red) and mesodermal cells (gray) migrate to the inside of the embryo. Neurulation (bottom) forms the neural fold on the dorsal side of the embryo, shown here as a cross section, and eventually gives rise to the neural tube.

At the beginning of gastrulation, cells migrate into the embryo to form the mesoderm. This creates a ring-shaped opening on the surface of the blastoderm, which is located just below the equator and is known as the blastopore. The cells on the dorsal side of the blastopore will develop into the notochord. The neighboring cells later give rise to musculature, vertebrae, and heart, while the cells from the ventral side develop into blood. A multilayered horizontal axis starts to develop on the future dorsal side of the blastopore in an area known as the dorsal lip. It is this very region that the German biologist Hans Spemann transplanted and identified as the organizer in his famous experiment (see Chapter II).

The mesoderm cells converge toward the dorsal side and migrate inward and upward in succession. The first cells to invaginate position themselves in front of the organizer and, together with the overlying ectoderm, will form the head structures. The mesoderm cells, that ingress later, squeeze in between those already present, and thereby elongate the axis. The ectoderm also simultaneously stretches downward and eventually engulfs cells of the bottom half. The large, yolk-rich cells develop into the endoderm, while the others form mesoderm. At this point, the embryo has become multilayered with an obvious anterior–posterior organization. The blastopore cells develop into the anus, the mouth forms anew, and the tail bud marks the posterior end (Figure 43).

Development progresses from front to rear. The notochord, the rod-like stabilizing structure, develops from the centrally invaginated cells. The somites develop in the mesoderm on either side of the embryo. The first somites appear at the front, and more somites subsequently form in regular intervals toward the back. Later, three different tissues will develop from the somites: the cartilage and bones of the vertebrae, the body musculature, and the inner layer of the skin. The outer skin, the epidermis, will develop from the ectoderm.

A process called neurulation creates the nervous system from the ectoderm. A groove forms in the ectoderm on the dorsal side above the notochord; this marks the beginning of the formation of the neural tube (Figure 43). At the head end, the neural tube widens into three larger vesicles that will form the forebrain, midbrain, and hindbrain. The neural tube itself will become the spinal cord that will be encased by the vertebrae. The vertebrae are built from both the somites and the notochord. The eyes form as bulges on the forebrain, while the lenses, nose, and ear structures develop from thickenings in the ectoderm known as placodes.

Special cells located over the dorsal neural tube form the neural crest. These cells migrate over large distances in the embryo toward the ventral side where they contribute to several structures. Among other things, the neural crest cells give rise to the bones of the head skeleton, the skull and the jaw bones, and the pigment cells that are responsible for the coloration of the skin. They also contribute to the innervation of several sensory organs. The neural crest is also responsible for a number of surface characteristics that distinguish vertebrate species from each other such as antlers, horns, head shapes, and color patterns.

Zebrafish and Chickens. The development of the fish embryo resembles that of the frog in many respects. There is, however, one big difference that concerns the separation of the embryonic cells and the yolk. In the fish, cleavage does not divide the entire cell content, but is confined to a clear zone containing the egg cytoplasm. After cleavage, the fish blastula consists of many small cells that begin to spread like a cap over the yolk. Once this cap of cells reaches the equator, the cells at the migrating edge of the cap begin to invaginate to form the mesoderm and the endoderm. The axis then develops on the dorsal side of the blastula beginning with a thickening known as the shield. This shield is the fish equivalent of the frog organizer. The intestinal tract will later encase the yolk, and its nutritional resources will eventually allow the embryo to grow.

The separation of egg cell and yolk is even more dramatic in bird embryos. The actual egg cell is the yellow yolk of the chicken egg. Fertilization takes place one day before the egg is laid in a small island of cytoplasm that contains the egg cell nucleus. Cleavage divisions then create the blastoderm, which is a round disc positioned on top of the yolk. Thus, the bird embryo develops almost two-dimensionally in a way that makes its anterior–posterior organization much easier to understand than that of the frog embryo.

The development of a longitudinal indentation, known as the primitive streak, stretches from the center to the tail end of the blastoderm. A comparison of the fate maps of frog and chicken embryos reveals that the chicken's primitive streak is homologous to the frog's blastopore (Figure 44). In the chicken, a structure known as Hensen's node develops at the frontal end of the primitive streak, which is the equivalent of the frog's organizer. At Hensen's node, cells ingress along the primitive streak and form the mesoderm and endoderm germ layers. But unlike in the frog and fish, the cells of the bird embryo already grow before they divide at this stage. Beginning at Hensen's node, the embryonic axis elongates toward the posterior end (Figure 45).

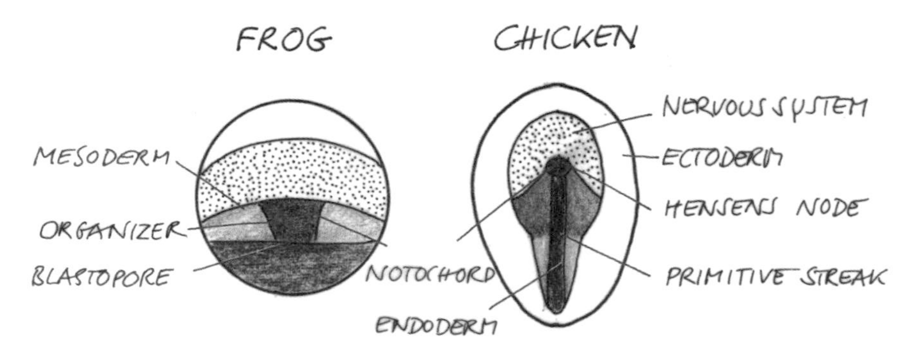

Figure 44. *Fate Maps of Frog and Chicken Embryos.* The two fate maps differ most regarding the location of the developing endoderm and mesoderm that migrate inward during gastrulation. The frog embryo's blastopore ring corresponds to the chicken's slit known as the primitive streak. The chicken's organizer is called Hensen's node. This figure has been redrawn from Lewis Wolpert's (et al.) book, *Principles of Development*. Oxford University Press, 2002.

In the case of birds, the embryo and its nutrients contained in the yolk are separated to an extreme degree. To make the nutrients available, cell layers spread above the yolk to form extraembryonic membranes and tissues. These will temporarily serve to nourish and protect the embryo. One such extraembryonic tissue is the yolk sac, which emerges from the endoderm and grows around the yolk mass. The yolk sac itself is covered by layers of mesoderm cells that will soon develop blood islands and blood vessels. These, in turn, will eventually connect to the embryonic vessels and transport nutrients from the yolk to the embryo. Another extraembryonic tissue is the amnion, which emerges from ectodermal folds and encloses the embryo to form a protective, fluid-filled cavity (Figure 46).

The first functioning organ of a chicken embryo is the heart—Aristotle's "salient point" as described previously in Chapter I. It develops from paired structures in the mesoderm on the ventral side, and, once formed, drives the blood circulation around the yolk long before the body's circulatory system has emerged. It is rather impressive that Aristotle already recognized that the yolk circulation in the chicken is similar to the workings of the placenta and umbilical cord in mammals. The chicken's allantois, a sac also containing blood vessels, transports oxygen to the embryo and removes waste products. It thus fulfills the functions of the lungs and the kidneys which will develop much later (Figure 46).

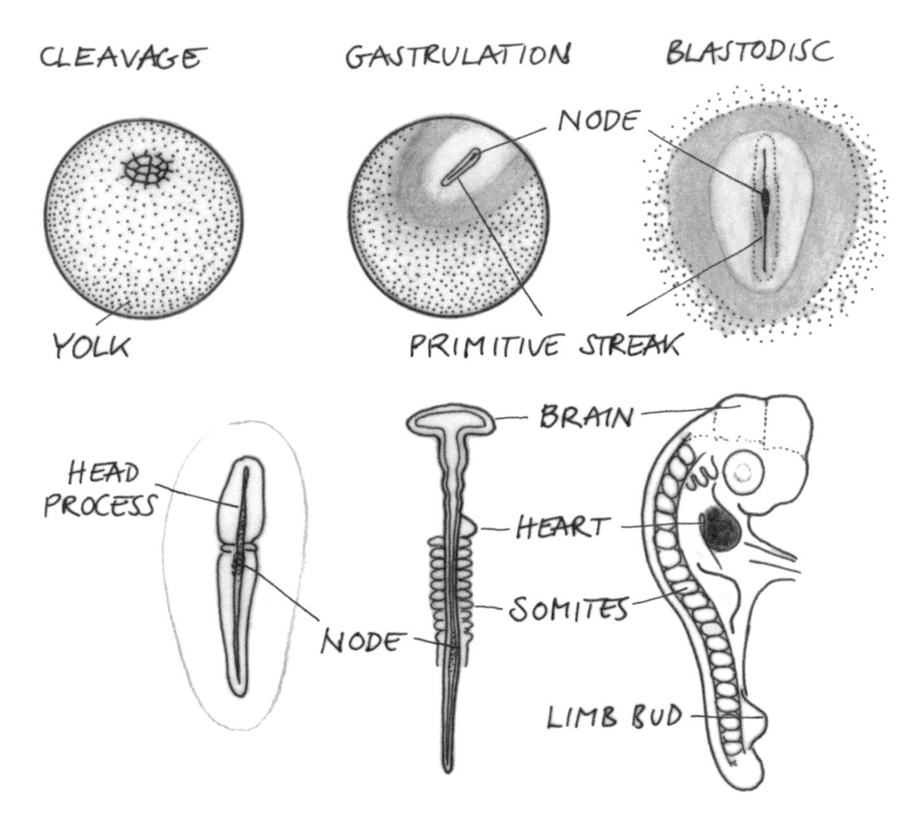

Figure 45. *Development of the Chicken: Formation of the Embryo.* The germ disc forms a horizontal groove, known as the primitive streak, which allows cells of the mesoderm and endoderm to migrate inward (see also Figure 46). A thin layer of cells covers the yolk at the edge of the disc. The anterior-posterior axis develops starting at Hensen's node. Along the inside, cells move to the front to form the head. The node progresses toward the back as the axis elongates. Somites form at the sides. The heart develops after two days in the chicken embryo, moving blood along the veins that cover the yolk in order to provide the embryo with nutrients.

2. *Mammals: The Mouse*

Frogs, birds, and fish lay eggs that contain all the nutrients and factors necessary for the development of an independent chicken, tadpole, or fish outside of the mother. All it takes for an egg to develop into an independent organism is a certain outside temperature, humidity, and air. This is not true, however, for mammals. Mammals give birth to live young; that is,

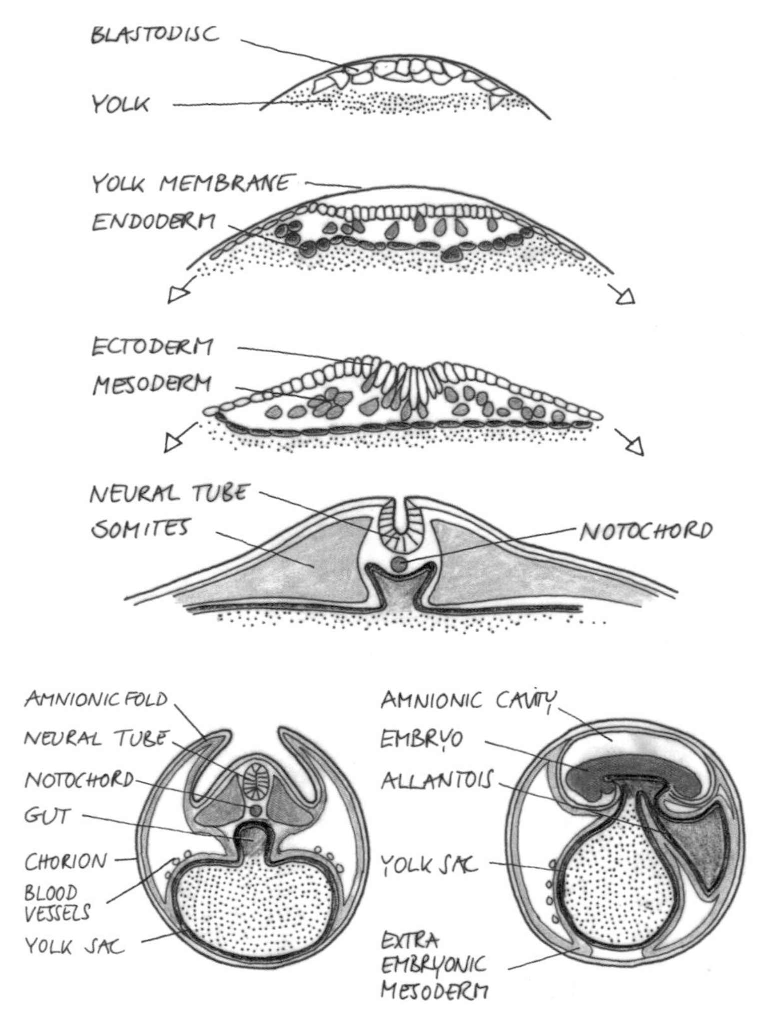

Figure 46. *Development of the Chicken: Gastrulation and Coverings.* The migration of cells into the primitive streak results in the embryo consisting of three layers that look like a flat disc lying on top of the yolk. Cell sheets of all three germ layers grow around the yolk to form the extra embryonic coverings (chorion, extra embryonic mesoderm, yolk sac, allantois, and amnion) that are important to nutritional intake and disposal. The yolk sac encloses the yolk. Blood cells and blood vessels are formed by the yolk sac, and connect to the embryonic heart early on during development. The allantois is an endodermal sac collecting waste products and exchanging oxygen. The amnion forms as an ectodermal fold closing over the dorsal side to protect the embryo. It stretches to the side to eventually cover the entire embryo and to bundle together the extra embryonic sacs and vessels ventrally. The figure shows the endoderm in red, the mesoderm in gray, and the yolk stippled.

embryonic development from fertilization to birth occurs within the body of the mother.

Mammalian development occurs in two phases. The mammalian egg contains only the information and material necessary for the development of a so-called blastocyst. The blastocyst does not contain much more than 100 cells, which give rise to both the embryo and part of the placenta, an extraembryonic structure. During the second phase, the blastocyst implants into the wall of the female uterus. Only during this phase, the embryo grows and takes on shape. In this regard, mammals differ from other animals as there is a time delay in the onset of the mother's care for the embryo relative to the time of fertilization. Egg-laying animals supply their future embryos with nutrition before fertilization, whereas mammals only do so after the embryo has been implanted in the uterine wall.

Development Inside the Egg. The eggs of mammals are some of the smallest eggs known. For example, the mouse egg has a diameter of only 80 micrometers. The egg is wrapped in a clear shell known as the zona pellucida. Fertilization takes place in the fallopian tube, where the egg swims in a body fluid. The cleavage divisions occur at a relatively slow pace of every 12 to 24 hours. The first embryonic cells are round and only loosely organized. Once eight cells are formed, they attach more closely to each other to create a rather compact cell ball known as the morula. This is achieved through the expression of cadherin cell adhesion molecules. After further divisions, when there are approximately 64 cells, a cavity develops giving rise to the blastocyst (Figure 47). The outer cells of the blastocyst form an epithelium called

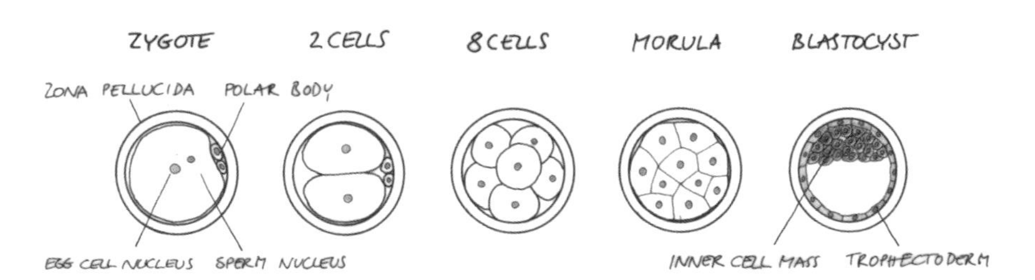

Figure 47. *Early Mouse Development.* The yolkless egg is enclosed by the zona pellucida. The polar bodies develop as a product of meiosis (see also Figure 51). After cleavage, the morula forms. At the blastocyst stage, the cells arrange into an inner group known as the inner cell mass (red) from which the embryo as well as extra embryonic tissue will form. The outer cells known as the trophectoderm will contribute to the placenta.

the trophectoderm, while the inner cells collect loosely on one side of the blastocyst to form the inner cell mass. From this stage on, the cells of the trophectoderm and the inner cell mass behave as separate groups, and the cells of one group can no longer contribute to the other. This differentiation into two cell types, one of which is extraembryonic, is only required for the attachment of the embryo to the uterine wall. The trophectoderm will contribute the chorion, which together with maternal tissue forms the placenta. This organ is an "invention" of mammals and provides the nourishment for the growing embryo by the maternal organism.

Experiments with mouse embryos have shown that, if separated, a single cell from a two- or four-cell stage embryo can still develop into a living mouse. This is no longer possible for individual cells from embryos of later stages, perhaps because they are just too small. It is possible to remove one or two cells from an eight-cell stage embryo without impairing its development. Mixing cells of two eight-cell stage embryos will create a mouse that displays characteristics of both embryos, but in each case the contributions from the original embryos will be different. This experiment demonstrates that, at this stage, the cells are not yet restricted in their developmental potential, but are still pluripotent, meaning that each cell can still become any differentiated cell type (Figure 48). Such animals developing from cells of two different embryos are called chimeras. Chimeras can also be created by a transplantation of cells from the inner cell mass of one embryo into the inner cell mass of another embryo. Also in this case, all the cells can still contribute to all tissues, even to the germ line. They apparently adapt fully to their new position in the recipient embryo. However, without the trophectoderm, the inner cell mass cannot give rise to a mouse because the implantation and placenta formation depend on the trophectoderm. These are functions that the cells of the inner cell mass cannot perform.

Imprinting. Imprinting is a phenomenon unique to mammals and concerns the differentiation between the trophectoderm and the inner cell mass. It is relevant whether certain genes that influence this differentiation are contributed by the sperm nucleus from the father or by the oocyte nucleus from the mother. Some maternal genes repress the formation of the trophectoderm cells while some paternal genes stimulate them. Both sets of genes in combination are necessary to guarantee a balance between trophectoderm and inner cell mass.

Imprinting is based on sex-specific activation of certain genes in the female and others in the male germ cells. This implies that in mammals, without contribution from the male sperm nucleus, it is not possible to initiate normal

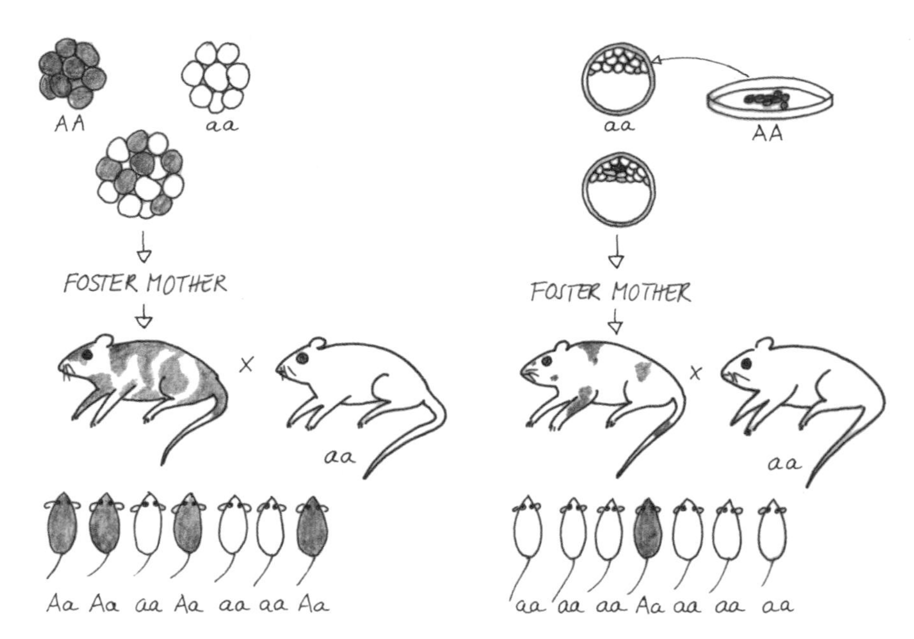

Figure 48. *Chimeras*. If two mouse embryos of different genotype are combined at the eight-cell stage and transplanted into a surrogate mother, they give rise to a chimeric mouse which consists of tissues derived from each of the donors. This can be apparent in the color of the fur. The figure on the right shows how single cells from a culture of embryonic stem cells are transplanted to a blastocyst. When the chimeras are crossed to mice of a given phenotype, some offspring will show the donor genotype A in all tissues. This indicates that the donor cells have given rise to germ cells in the chimera.

embryonic development from an unfertilized egg by parthenogenesis. In contrast, in the frog and fish, for example, parthenogenesis may be achieved if nuclei of the polar bodies, instead of a sperm nucleus, can be made to fuse with the oocyte nucleus. This results in a diploid egg cell that can develop normally, albeit with low efficiency. This is not possible in mice, because embryos exclusively derived from female egg cells are deficient in trophectoderm, the development of which demands gene activity from the male sperm cell.

The Development in the Uterus. The blastocyst hatches from the zona pellicula and attaches to the extracellular matrix of the uterus mucous membrane. The trophectoderm cells then embed themselves and the blastocyst until the blastocyst is completely wrapped with uterine tissue. The trophectoderm cells divide and penetrate the surrounding tissue. Eventually, together with uterine tissue, they form the placenta that is nourished by maternal blood vessels. Later, the umbilical cord will connect the system of embryonic blood vessels with the placenta so that the mother's blood circulates around the embryo's blood vessels. The umbilical cord guarantees the supply of oxygen and nutrients and provides the passageway for waste products. In this way, the embryo makes use of the mother's organs as long as its own organs cannot function independently.

After implantation, the embryo and the extraembryonic layers, such as the amnion, yolk sac, and allantois, emerge from the inner cell mass. The embryo itself develops from an epithelial structure called egg cylinder. The egg cylinder is comparable to the early embryo of the chicken, but its shape is more like an indented cup-like structure. As in the chicken, the axis of the mammalian embryo develops by the formation of a primitive streak and a node that is homologous to the chicken's organizer. The streak and node then move toward the bottom as elongation of the axis occurs. The final arrangement of the embryonic structures is reached later in a remarkable process, during which the embryo twists itself lengthwise on its own axis causing its inside to be turned out, and is wrapped up in the amnion.

Embryonic Stem Cells of the Mouse. In certain culture media, the inner blastocyst cells can be multiplied without changing their undifferentiated condition (Figure 40). These inner blastocyst cells are called embryonic stem (ES) cells. Embryonic stem cell cultures were first established from mouse blastocysts in the 1980s. If embryonic stem cells are transplanted into a host embryo, they will participate in its development exactly as the cells of the host embryo. The offspring of these foreign cells can contribute to every tissue, even to the gametes (Figure 48). In a Petri dish, these embryonic stem cells may form a random assortment of different cell types, such as nerve, muscle, and blood cells. Yet, they cannot organize themselves to actually build an organism. In other words, a mouse cannot form in a Petri dish. For a mouse to develop, it is absolutely necessary to have the organizing influences of both the outer trophectoderm cells of the blastocyst and of the mother's uterus after implantation.

The specific qualities of embryonic stem cells offer unique opportunities for research. For example, the development of particular cell types can be stimulated or repressed in a cell culture by adding certain growth factors. In the case of the mouse, such in vitro differentiated cells can be transplanted into an adult organism, where they sometimes integrate into the respective tissue. Such experiments are not only relevant for basic science, but also for the development of therapies based on cell replacement.

Since embryonic stem cells grow well in culture, it is possible to conduct experiments that involve rare events such as the introduction of an additional gene into the cell's genome. An isolated gene, when added to the cell culture, may integrate into the genome of an embryonic stem cell. Cells that are transformed in this way can be selected on the basis of whether they have integrated the additional gene into their genome or not. These genetically altered cells can be transplanted into a blastocyst where they contribute to the development of a mouse (Figure 48). Some of the cells of the resulting mouse and, in rare cases, even its germ-line cells, can be traced back to these transformed stem cells. The offspring derived from these germ-line cells will carry the foreign gene in all of their cells. These transgenic animals (see also Chapter III) will pass the foreign gene on to the next generation just like their own genes.

Embryonic stem cell cultures allow for the introduction of mutations in genes of the mouse. For example, when DNA of a specific mutant gene is added in vitro to the embryonic stem cell culture, it may replace the chromosomal gene, a process known as homologous recombination. During this process, a crossover exchange of genomic DNA for plasmid DNA takes place at the corresponding sites in the genome resulting in the creation of embryonic stem cells that are mutant for the respective gene. Using a specific culture condition, these mutant cells can be selectively grown and isolated. If they are transplanted into a blastocyst, after one further generation, eventually a mouse carrying only the mutant version of the gene may emerge (Figure 48). The mutant mice that develop in this way are known as knockout mice (Figure 49). Homologous recombination is used mainly to exchange an intact gene with one that is mutant, in order to shut down or alter the function of the gene and study the consequences this has on the mouse. These knockout mice are an important tool for studying the function of genes of known structures in the organism.

3. Gradients, Prepatterns, and Induction in Vertebrates

The spatial changes during the emergence of body shape are preceded by molecular prepatterns also present in vertebrates. As in *Drosophila*, gradients determine polarities and patterns, as well as overlapping distributions of transcription factors that form temporary prepatterns and eventually determine the final shape of the embryo. These processes are similar to those already described for *Drosophila*, though the individual steps in vertebrates are not as clearly separated as in the fly. In addition, vertebrate cells form membranes right from the start of development, so the spreading of signals must

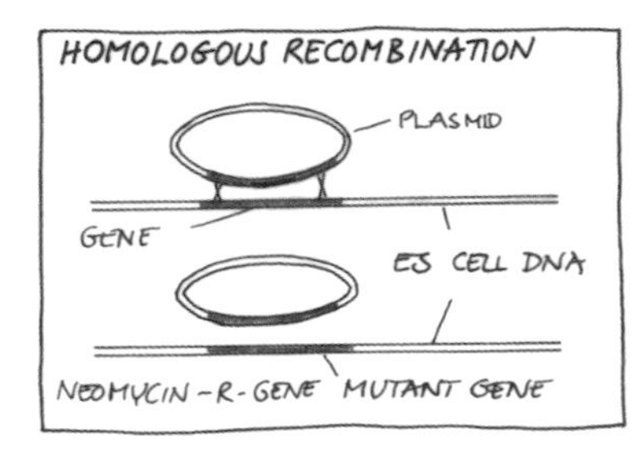

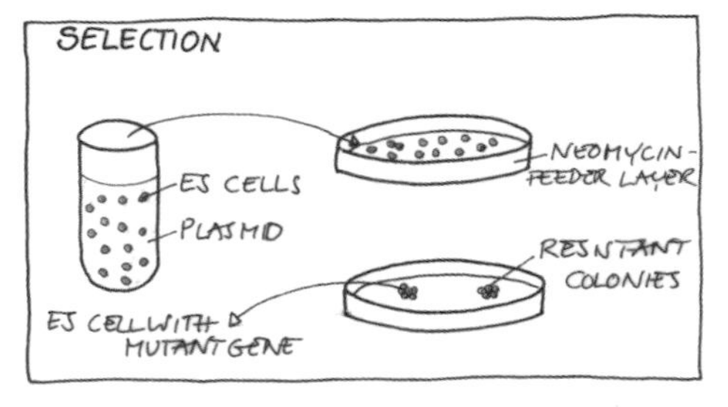

Figure 49. *Knock out Mice.* A mutant of a specific gene can be created by first generating mutant mouse embryonic stem cells in culture (left). A plasmid with the desired gene is transferred into the cells. The plasmid contains a copy of the gene to be inactivated. This copy has parts of the gene replaced with a gene causing resistance to an antibiotic. This insertion renders the plasmid copy of the gene of interest nonfunctional. In rare cases, a crossover between plasmid DNA and the corresponding chromosomal DNA takes place through a process called homologous recombination. Through this process the intact gene of the embryonic stem cell is replaced by the defective gene from the plasmid. Only cells in which such an event has happened will be able to survive in the presence of the antibiotic (right panel). Therefore the few cells in which the gene has been replaced can be selected easily. In order to create mutant mice, the embryonic stem cells with the mutant gene have to be transferred into mouse blastocysts as shown in Figure 48. After two generations of inbreeding, also individuals that are homozygous mutant for this gene will occur.

take place by communication between cells as opposed to the simpler diffusion occurring in the early fly embryo. In addition, the vertebrate blastula consists of several cell layers, and its massive cell movements and form-building processes already begin when the molecular prepatterns are still arising in other parts of the embryo. Therefore, the relationship between prepatterns and the actual forms are better defined in *Drosophila* than in vertebrates.

Gastrulation. Maternal factors, mainly asymmetrically localized mRNAs, determine the top-to-bottom axis in frog and fish eggs. Starting from these localized mRNAs, diffusible factors determine the endoderm and induce formation of the mesoderm in a neighboring belt-shaped region. A signal determines a specific region that forms the organizer on the future dorsal side. Molecules of the wingless family of signaling molecules then lead to an activation of several genes within this organizer. Some of them encode transcription factors, such as goosecoid, that in turn activate even more genes.

Organizer molecules such as chordin have long-range effects on their environment. Chordin is involved in building a gradient that determines the dorso-ventral axis. It does so by repressing the activity of a morphogen, the BMP protein, which is closely related to the Dpp protein of *Drosophila*. The organizer, thus, is not an inducer itself but an inhibitor of the BMP-inducing signal. The result is a BMP morphogen gradient with a maximum concentration on the future ventral side. High BMP concentration results in the formation of ventral structures such as blood and kidneys while repressing the development of the nervous system. Lower BMP concentrations induce mesodermal structures such as the somites, and the lack of BMP permits the nervous system to develop. The cells of the organizer form the notochord.

Several transcription factors are activated simultaneously in the cells of the blastopore, which will build the mesoderm structures. Among them is the *T*-gene (T stands for tail), which had been discovered in the mouse based on its dominant short-tail phenotype; it is called *brachyury* (short tail) in frogs and no tail in fish. At first, the *T*-gene is transcribed in all invaginating cells of the blastopore and then is later limited to the cells that form the notochord (Figure 50). In the case of the mouse, the *T*-gene is transcribed first in the primitive streak and then in the developing notochord.

Neurulation. The ordered ingression of mesoderm cells occurs during gastrulation as soon as the dorsal side of the embryo has been established. The anterior–posterior axis forms in the three germ layers with the notochord at the center. The axis is laterally confined by the precursors of the somites and covered by ectodermal tissue. Signals from the organizer induce the central nervous system in the overlying ectoderm. During this process, the most important function of the organizer is the suppression of BMP, which prevents the development of nervous tissue in favor of ectoderm.

These molecular findings finally explained many historic experiments. For example, if frog blastomeres are separated after the first cleavage division, both the right and left cells receive part of the organizer and can therefore still form a frog, albeit a smaller one. Artificial separation into a front half and a back half gives two cells, only one of which contains the organizer, thus leaving the other cell to form only a so-called bellypiece. The transplantation of the organizer, as in Hans Spemann's experiment, results in the production of the chordin protein on the ventral side, which represses the BMP protein. Therefore, in the neighboring tissue an additional axis is induced.

Segmentation. The segmental organization of the vertebrate body is not as apparent from the outside as it is in arthropods, but it is obvious in the body

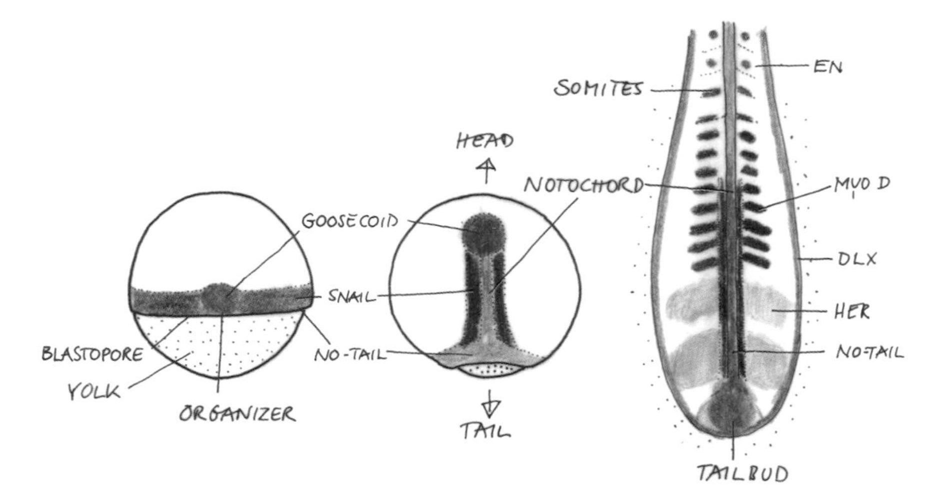

Figure 50. *Molecular Prepatterns in the Zebrafish Embryo.* During early development, a number of genes are active in the blastopore. On the left, looking at the organizer, the product of the no-tail-gene (gray), of the snail-gene (red), and the goosecoid-gene (black) are depicted. During the development of the body axis, the areas in which these genes are active separate (middle panel). The no-tail product is present in the future notochord, goosecoid at the frontal tip, and snail at the side of the mesoderm where the somites will be formed. Prepatterns during segmentation are shown on the right. The segmental organization of the muscle-forming region is revealed in the pattern of transcription of the myoD-gene (red). This is a result of transcription of the her1-gene in waves that start at the tailbud. They are shown in gray. The no-tail-gene (black) continues to be active in the notochord region.

musculature and skeleton of the vertebrate spinal column. Both musculature and vertebrae develop from the mesoderm, which itself is divided into regular blocks, the somites. In vertebrates, somites arise sequentially from anterior to posterior. Segmentation is caused by the timed pulsation of transcription of certain genes, for example the her-gene, in the tail bud. The rhythmic switching on and off of gene expression propagates like a wave toward the anterior of the germ band. It determines where a somite boundary will form when the wave encounters the front of the somites that have already formed. The rhythmic gene expression is the result of a complex interaction between proteins of the Delta–Notch group. Segmentation can be visualized as a prepattern of the transcript of the *myoD*-gene in the front margins of the somites (Figure 50).

Differentiation of the somites on either side of the notochord is influenced by the signaling protein product of the sonic hedgehog-gene that originates in the notochord. The same signaling system influences the organization of the nervous system located just above the notochord. For example, it induces the formation of motor neurons in the spinal cord.

Hox-Genes. The Hox-genes determine a variety of structures along the anterior–posterior axis. Vertebrates have 13 different Hox-genes organized into one gene complex. There are four such Hox-gene complexes in vertebrates, located on different chromosomes. Just like in *Drosophila*, the Hox-genes work in combination (see Chapter V). Hox-genes are activated along the anterior–posterior axis of the embryo, such that Hox-genes on one end of the complex are transcribed in anterior regions and Hox-genes further along the complex are transcribed in more posterior regions of the body. The specific combination of Hox-genes that is active at any given position along the embryonic axis influences among other things the shape of the vertebrae and the location of the ribs. They also determine the position of front and hind limbs of the embryo.

Formation of Limbs. The limbs develop from buds formed by an aggregation of mesoderm cells located underneath the skin. The growth factor FGF initiates the emergence of the limbs from the tip of these buds, while the signaling protein sonic hedgehog determines their pattern along the anterior–posterior axis of the limb. After the limbs emerge, mesenchyme cells in the limb bud then condense to form cartilage tissue. Later, this cartilage develops into the bones of the animal's front and hind limbs. Fingers and toes develop from plates of cartilage tissue by controlled death of cells in rows between the emerging digits.

CHAPTER VIII

Humans

HUMANS OCCUPY A SPECIAL POSITION WITHIN BIOLOGICAL RESEARCH. ON THE one hand, medicine has collected an astonishing amount of data on diseases, the immune system, physiology, and nutrition. Moreover, much information on general biochemistry and cell biology has been obtained by studying cultured cell lines established from human cancers or connective tissue. On the other hand, the understanding of gene function in the living human organism is rather incomplete. This is due to the fact that the genetic experiments performed on animals are ethically impossible to do on humans. Therefore, our knowledge of human genetics comes to a large extent from "natural experiments" such as diseases, accidents, spontaneous mutations, and observations during therapy.

Proper biological experiments, however, always include control groups, a large number of test subjects and reproducible conditions. For example, genetic experiments involve breeding of carefully selected parents, mutagenesis, inbreeding, and genetic transformation, all of which can be easily performed on flies, mice, and fish. Genetic studies of this kind have provided most of our knowledge about genes.

As none of these methods can be applied to humans, human gene function must be largely inferred from research on model organisms. In a biological sense, humans are, of course, nothing more than rather sophisticated animals. Indeed, the genome of the mouse is very similar to that of humans. However, other model organisms, such as rats, rabbits, pigs, dogs, and cats, are even better suited for certain kinds of experiments relevant to human biology. Furthermore, there is a large degree of homology between the genes of different organisms and the respective gene can be investigated in great detail in a fly and compared to its counterpart in a fish or mouse. One might, for example, search for genes with similar functions in the model organisms and subsequently investigate their relevance to human biology. In this way, research on model organisms is not only advancing general knowledge, but also yielding results that are immediately relevant for human biology, as they could never be obtained directly from studies on human subjects. In other words, it is much easier to ask the question: "is gene X doing the same in humans as in flies or mice" than to investigate the problem directly in humans.

1. The Development of Germ Cells

Boy or Girl? In mammals, the question is decided by the type of sex chromosomes present in the zygote. If two X-chromosomes are present, the resulting organism will become a female, if one X- and one Y-chromosome are present, it will become a male. In that sense, the presence of a Y-chromosome determines which sex will be formed, or more precisely, the presence of the sry-gene, which is located on the Y-chromosome. If a Y-chromosome, and thus the sry-gene, is present, the gonad precursors will eventually become testes; if not, they will develop into ovaries. In rare cases, the sry-gene happens to be transferred to an X-chromosome and the resulting child will develop male gonads, even though it has two Xs.

During embryonic development, the sry-gene instructs the male gonads to produce the male hormone testosterone. Testosterone then spreads through the bloodstream and results in the formation of male characteristics in other parts of the body. If testosterone is absent, the female gonad will form other hormones that trigger the formation of female characteristics. This means that only one gene, the sry-gene, makes the difference between male and female gonads, but the sex-specific characteristics of the rest of the body known as secondary sexual characteristics are due to several hormones.

Other than the sry-gene and very few other genes on the Y-chromosome that are necessary to produce sperm cells, both sexes have the same set of genes. As mentioned above, girls have two X-chromosomes, boys only one. This would mean that girls get twice the amount of gene products from genes on the X-chromosome. Therefore, the genetic activity of the X-chromosomes must be balanced in males and females. To this end, early in female development, one of the two X-chromosomes is shut down in the somatic cells—which one is left to chance. This ensures that boys and girls receive the same dosage of gene products from the X-chromosome.

The germ cells, which will later differentiate into eggs or sperm, develop from a group of about 50 cells originating in the yolk sac. During gastrulation, these cells migrate into the embryo and associate with the gonads. While all prospective egg cells are created during the female's embryonic development, sperm cells are produced continuously after a male has reached puberty.

Egg cells are precious. Only about 400 of them mature during a woman's life. And even though mammalian eggs do not contain yolk, they are still about 100 times larger than the average cell in the body. With their large size, they are endowed with enough cytoplasm necessary for the formation of a blastocyst of approximately 100 cells. In contrast, males can produce millions of sperm cells containing essentially only a nucleus, a flagellum for locomotion, and a few mitochondria.

Meiosis. Germ-line cells are diploid at first, containing all 23 human chromosomes twice, with the exception that male germ-line cells contain one X- and one Y-chromosome, rather than two X-chromosomes. The mature germ cells, egg cell and sperm, are haploid and contain only one copy of each chromosome; in males, they contain either an X- or a Y-chromosome, which is why the sex chromosome of the sperm determines the sex of the fertilized oocyte, as the female can only contribute X-chromosomes. The process leading from the diploid to the haploid state is called meiosis. After chromosome duplication, two divisions halve the number of chromosomes. The daughter cells of the first meiotic division contain one copy of each chromosome, which, however, has already been duplicated and recombined. The second division is a normal mitosis during which the chromatids are pulled apart (see Figure 39).

Meiosis is a complex process during which errors may result in defective distribution or breaks of chromosomes. Many recombination errors are noticed by repair enzymes, and defective germ cells are eliminated by apoptosis. Sometimes, however, more frequently with a woman's increasing age, defective egg cells containing too few or too many chromosomes, may mature. Such chromosomal disturbances, known as aneuploidies, are the cause of many miscarriages and even sterility in humans. When mistakes have happened during the separation of homologous chromosomes or chromatids, the nucleus of an egg cell may receive two copies of a chromosome, or none. In such cases, fertilization with a haploid sperm will result in an embryo with three copies (trisomic) or only one copy (monosomic) of the respective chromosome. All monosomies are lethal at a certain time before or shortly after birth—all, except, of course, the X-chromosome, as males have only one X. Trisomies are also lethal and in most cases the embryo stops developing before birth. One exception is the trisomy of chromosome 21, which leads to Down syndrome. Chromosome 21 is one of the smallest human chromosomes, containing an unusually low number of genes; hence, the organism can tolerate the fact that there are three instead of two copies. Nevertheless, children with Down syndrome exhibit cognitive disabilities and physical abnormalities even before birth.

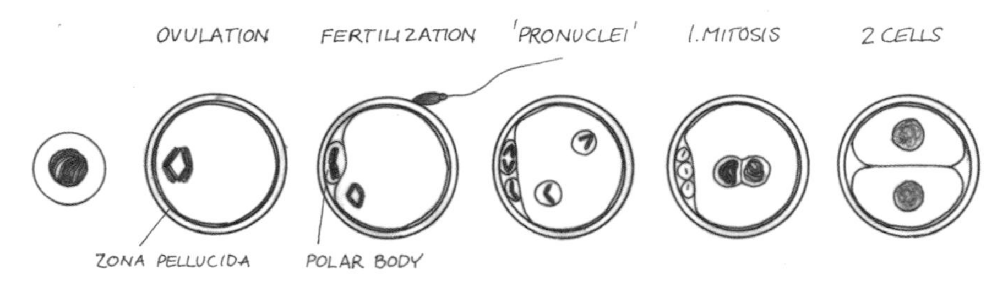

Figure 51. *Fertilization in Mammals.* During the maturation of the egg only one of the four haploid products of meiosis becomes an egg cell. All others are split off as polar bodies (see also Figure 39). During fertilization the egg is still undergoing meiosis. As soon as the sperm and egg pronuclei have met, the chromosomes are doubled, and in turn this initiates the first cell division. This figure shows only one chromosome pair.

Sperm and Egg Cells. The testicles of newborn boys contain a supply of diploid sperm stem cells. Actual sperm cells begin to develop in puberty. The stem cells divide in a typical asymmetric stem cell division (see Figure 40), which creates a stem cell and a sperm mother cell that continues to divide. After the two meiotic divisions, four haploid sperm cells emerge from each sperm mother cell (see Figure 39). This process continues throughout a man's life. With egg cells, however, things are different. The ovaries of a newborn girl already contain thousands of immature egg cells, which have been arrested during the first meiotic division. The stem cell divisions have already been completed before birth. After puberty, one egg matures in the ovaries per month. The egg is surrounded by many supportive cells that form a follicle. During maturation, the egg cell grows from 10 to about 100 micrometers in diameter. A transparent shell known as the zona pellucida and consisting of extracellular matrix, covers the egg. After the first meiotic division is completed, one of the daughter nuclei of this division, containing a complete set of chromosomes, is extruded from the egg as a so-called polar body, which eventually will disappear (Figure 51). As soon as the egg cell is ready for fertilization, it is pushed from the follicle into the fallopian tubes. This process is called ovulation.

2. *The Development of the Egg*

Fertilization and Formation of the Blastocyst. Fertilization takes place in the fallopian tube, as the sperm's head penetrates the zona pellucida. It is only at this stage that the second polar body is pushed out and meiosis is completed.

Inside the egg, the nuclei of egg and sperm called pronuclei migrate toward each other. As soon as they meet, their chromosomes double, thus initiating the first cell division of the embryo. In humans, unlike in many other organisms, the pronuclei do not fuse, instead they divide separately. At the two-cell stage, the maternal and paternal chromosomes are combined in one nucleus per cell for the first time (Figure 51). After cleavage, compaction, and further divisions, the blastocyst develops, much like in mice (see also Figure 47). The blastocyst is a hollow ball of flat trophectoderm cells surrounding a cavity, which houses the inner cell mass. From this inner cell mass, the embryo as well as the extraembryonic tissues such as the yolk sac and the amnion will develop.

Implantation. During cleavage, the egg moves within the fallopian tube toward the uterus. Once it arrives, the blastocyst hatches out of the zona pellucida. Several components of the extracellular matrix, the uterine wall, and the surface of the blastocyst itself assist in attaching the blastocyst to the uterine wall (Figure 52). The uterus is hormonally prepared for this process of implantation, which takes place at about the fifth day after fertilization.

3. The Development in the Uterus

Placenta. Lodging into the uterine tissue requires the help of embryonic enzymes that dissolve components of the extracellular matrix of the uterine mucous membrane. The trophectoderm cells grow into the uterine lining and form the syncytiotrophoblast, a structure that emerges through cell fusion. The syncytiotrophoblast will later join with cells of the extraembryonic mesoderm to develop into the chorion. The chorion covers the embryo and deeply penetrates the mother's uterine tissue by many branched extensions. The tissue of the chorion and cells of the uterine mucous membrane form the placenta. In the uterus, cavities emerge, which are flooded with blood from the mother's blood vessels. In the chorion, embryonic capillary blood vessels develop, which spread in the cavities, such that they are bathed in the mother's blood. This permits exchange of molecules between the mother and embryo (Figure 52) without direct mixing. The mother's blood delivers nutrients, oxygen, and antibodies to the embryo and removes waste products. Signals between mother and embryo are constantly being exchanged in the placenta: embryonic factors initiate the blood supply of the uterine wall, while maternal factors stimulate the embryo's growth. If these latter signals cease, the embryo will die.

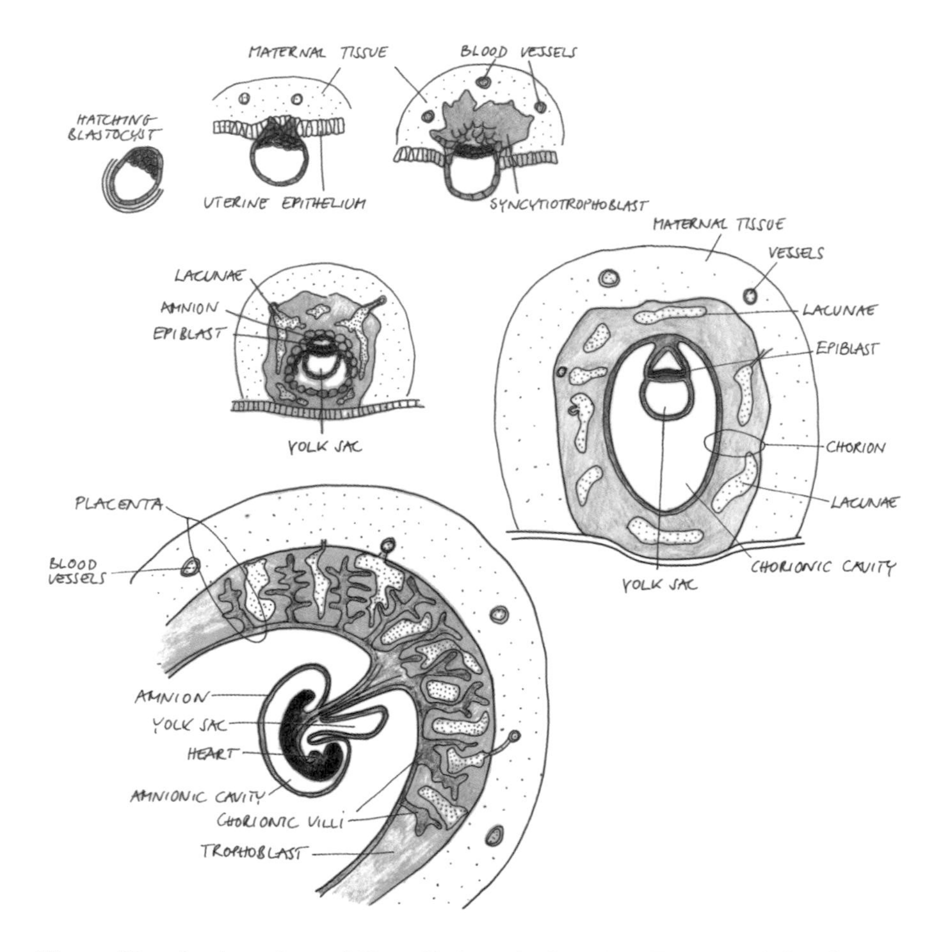

Figure 52. *Implantation and Extra Embryonic Coverings.* From top left to bottom right: the blastocyst hatches from the zona pellucida. It adheres to the uterine wall and implants. The trophoblast cells form a syncytiotrophoblast. The cells of the inner cell mass (red) form the embryo as well as the extra embryonic amnion and the yolk sac. After implantation, the chorion grows to form the chorionic cavity. The embryo is connected by a stalk to the chorion, which is covered with embryonic tissue on the inside. The actual embryo develops from the epiblast where just like with the chicken embryo, a primitive groove develops and initiates gastrulation. Later, the amnion stretches over the embryo and bundles the yolk sac with blood vessels to form the umbilical cord. Tissues derived from the inner cell mass are shown in red, those derived from trophectoderm cells in gray, and maternal tissue is shown stippled.

The Formation of the Embryo. After implantation, the inner cell mass grows and shapes into a two-layered structure resembling the disc-shaped structure of the chicken embryo. The cells of the lower layer spread to cover the inside wall of the trophectoderm, which corresponds to the yolk sac of the chicken egg. Starting from the upper layer, the epiblast, a membrane known as amnion develops and eventually encloses the fluid-filled cavity above the embryo. It is the epiblast that will give rise to the actual embryo. It develops a primitive streak and initiates gastrulation. After two more weeks, the germ layers have formed and the somites, the central nervous system, as well as the head, become apparent and the heart begins to beat. The amnion stretches around the embryo, enclosing it completely. During this process, yolk sac and blood vessels are bundled to form the umbilical cord, which maintains the contact of the embryonic blood vessels to the uterine wall (Figure 52). At this stage, about 5 weeks after fertilization, the embryo has developed most of its structures, even though it is still tiny, no more than a few millimeters long. Further development mainly entails growth and final differentiation of the embryonic structures.

It is important to note that also harmful influences can reach the embryo via the mother's blood, particularly during the early stages of the pregnancy when the organs and tissues of the embryo develop. Indeed, the morning sickness suffered by many pregnant women may be a defense mechanism against potential poisoning of the embryo. Sometimes, medication can harm the embryo, such as in the case of Thalidomide (Contergan), which impairs the formation of limbs, but only when ingested at the time during which limbs are formed between weeks 4 and 5. By contrast, drugs like nicotine, alcohol, or cocaine can have an effect throughout pregnancy, because they affect the formation and function of the embryonic brain, which is developing continuously. Although the barriers between maternal and embryonic blood do not allow larger molecules to pass, bleeding in the placenta may transfer viruses from the mother to the child. The embryo does not yet have a functional immune system and so the mother's antibodies enter the embryonic blood for protection.

Twins. As mentioned above, experiments that could tell us something about the developmental potential during the early stages of the pregnancy cannot be performed on humans. Yet a study of a natural phenomenon, twins, offers a few insights into these early developmental decisions. Fraternal twins, who are genetically as different as other siblings, arise when two egg cells, rather than just one, mature and are fertilized. They are separated from the very beginning and each forms their own chorion. Monozygotic twins, which are genetically identical, arise by splitting of a blastocyst derived from one fertilized egg cell. From the embryonic coverings, chorion and amnion in monozygotic twins, it is possible to deduce the time at which the splitting of the embryo has taken place. Some identical twins are each equipped with an individual chorion, indicating that division took place at early cleavage stages. Others share a chorion, which means that the division took place after the formation of the blastocyst. The latest stage at which one embryo can still give rise to twins—although this happens rarely and sometimes leads to joined body parts, so-called Siamese twins—is at about 14 days after fertilization, when the amnion has formed and gastrulation is about to begin (Figure 53). One could even argue that up to this point there is no individual embryo because the epiblast can still form two individuals. In other words, at least up to this stage, the cells are not determined to their future fate.

Twin studies also offer an insight into the question: which of our properties are determined by genetic factors and which by the environment; in other words, is it nature or nurture? Identical twins resemble each other much more than regular siblings—they are genetically identical and experience the same environment in the uterus of the mother. When twins are se-

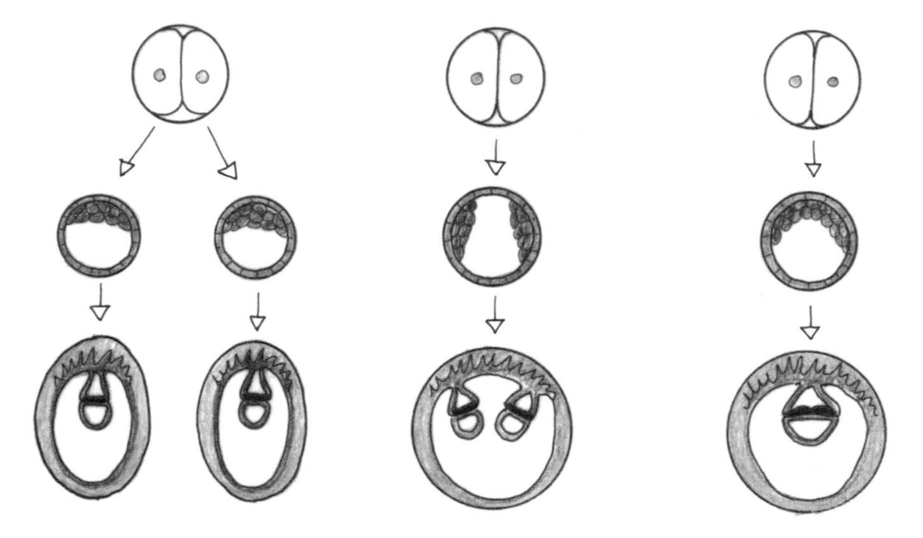

Figure 53. *Identical Twins.* Identical twins are a result of a splitting of the embryo at different stages of development. Early cell separation results in two smaller blastocysts that can develop normally (left). Sometimes, the inner cell mass divides, which results in two separate embryos sharing one chorion (center). When separation does not take place until the epiblast stage, two embryos develop sharing an amnion (right). If the separation is not complete, Siamese twins will develop.

parated after birth, the arising differences will be largely due to differences in social and personal experiences, nurture as it were.

Birth. During intrauterine development, the connection between embryo and mother is so intensive and strong that a separation would harm both the embryo and the mother. An embryo can survive outside the mother's body only for a short period of time—if at all. Compared to other closely related mammals, humans are born at a relatively early stage of their development. To reach the same degree of maturity as other mammals, a pregnancy of several more months would be required. This early birth can be explained partly by the exceptionally large size of the head, which keeps on growing continuously after birth. A child born later than 9 months of pregnancy would pose a serious threat to the mother's health. This early birth of humans has important consequences. For example, foals can walk and see right after they are born, but a human newborn is, in many regards, not quite as finished. Several features that develop before birth in some other animals, take place in humans only after birth.

Generally, the formation of sensory organs and nerve connections is not just continued after birth, but depends to a large extent on stimuli encountered outside the mother. Initially, many nerve connections are formed, but only those that are actually used and activated will be retained. If there are no sensory stimuli during the sensitive developmental periods, the relevant nerve connections will be missing forever and later activities cannot correct these early deficiencies. Observations of children that were neglected during this early developmental period confirm that outside stimuli are indispensable for the normal development of human cognitive capacities. Language and a long period of youth, endowed with a particularly great capacity for learning are peculiar to humans. Language itself is not innate, yet the ability to learn a language is. It is a capacity that is most active during a child's toddler years.

4. Genes and Diseases

There are obviously many hereditary traits that appear in the father or mother and the children of a family such as blemishes or particularities or special talents. Among them are, for instance, fused eyebrows, protruding chin, early graying, perfect pitch hearing, and many more. These traits can be explained by spontaneous mutations that arose perhaps hundreds of generations ago in an ancestor and persisted since then. Such mutations can emerge anew at any time and some of them may result in fatal diseases.

Congenital Diseases. Diseases resulting from a mutation in one specific gene (monogenic diseases) are relatively rare. They are passed on from generation to generation according to Mendel's laws. Mutations in many genes may never become apparent, either because the mutation does not affect the function of the gene, or because the mutant gene may cause death at early embryonic stages. Known recessive hereditary diseases usually manifest themselves at birth at the latest. The affected children often die within the first few years. By contrast, dominant mutations are passed on only when the effect is mild or when the disease becomes apparent only after affected individuals have had children. Such diseases often do not appear until later in life, such as Huntington's disease or hereditary forms of dementia.

Recessive hereditary diseases can only manifest themselves when both parents are heterozygous for a mutation of the same gene; this can be caused either by independent mutations or by the tendency to marry individuals from the same social group who often can be genetically related. When both parents are heterozygotes, the chance is 25% that the zygote carries two mutant alleles, one from the mother and the other from the father. In Central Europe, a rather common recessive hereditary disease is cystic fibrosis. About 1 in 22 adults carry a mutant allele of the cystic fibrosis-gene, which in the homozygous condition leads to death during childhood in most cases. Notably, most carriers have the same mutant allele, which implies that they share a common ancestor in which this mutation occurred.

Recessive illnesses linked to the X-chromosome are a special case, as they appear dominant in boys because these have only one X-chromosome. These illnesses, among them hemophilia and muscular dystrophy, are passed on to their male offspring by heterozygous mothers who are not themselves afflicted with the disease, as they have two X-chromosomes, one of which carries the normal allele of the disease gene (see pedigree X-chromosome; Figure 11).

The severity of monogenic hereditary diseases can vary from case to case. This is because individual humans, even if they carry the same disease allele, differ with respect to other genes. In addition, numerous environmental factors may influence the course of a disease. The function of many genes underlying hereditary diseases is now understood at a molecular level. Studies on affected families have made it possible to localize the mutations and eventually to clone these genes. This allows a genetic diagnosis of the disease before symptoms become apparent. Sometimes the molecular structure of the protein product immediately offers an explanation of the symptoms of a given disease. Knowing their molecular origin has made a few hereditary diseases curable. For most of them, however, there is no therapy yet, nor will there be one in the foreseeable future.

Genetic Disposition. Apart from diseases that can be traced back to mutation of a single gene, genetic disposition plays a role with several common human diseases. This holds true for certain kinds of cancer, age-related dementia, metabolic disorders, and psychological illnesses such as depression or schizophrenia. It can be assumed that many genes are involved in these diseases in one way or the other—genes that interact in complex ways so that their effects are not yet understood and perhaps may never be in all details. Therefore, such diseases are not as predictable as disorders resulting from monogenic mutations. Severe monogenic diseases emerge with a 100% probability in identical twins, whereas other diseases that are only enhanced by a genetic disposition emerge in both of them only with a low probability. In these cases, the influence of life style and environmental factors is equally important as the hereditary disposition.

Genes and Cancer. Cancer also results from gene mutations. In this case, however, the mutations arise newly in body cells during the life of an individual. To some extent, cancer is an unpredictable accident that cannot be prevented with certainty.

In a normal body, the somatic cells have a limited life span, as their capacity to divide is restricted. For example, cells of the human connective tissue can divide in culture no more than 50 times. Other cells may divide even less frequently. Within a tissue, cells are continuously under influences that control their future: mistakes during DNA replication initiate apoptosis and cell divisions take place only when initiated by a growth factor, while additional cell divisions may be blocked by other factors. In other words, a human tissue is a well-balanced system, in which individual cells adapt to their environment through a continuous interaction and exchange with their neighbors. Cancer cells, however, have lost their ability to obey these influences. They are characterized by two special qualities, both of which can be fatal to the organism. First, they no longer react to mechanisms controlling growth and cell division, but rather continue to divide. Second, they do not remain with their original cell group, but can dissociate from it and migrate into new, healthy tissue.

Tumors develop in the body as a result of mutations in certain genes that are passed on to the daughter cells as they divide. Therefore, a tumor stems from one single mutant cell, which then forms a clone of cells that all carry this mutation. Only in rare cases, a single mutation is sufficient to cause a tumor; generally, cells of a grown tumor have mutations in several, perhaps six or seven genes. This indicates that the development of tumors is a stepwise process: the first mutation in one particular cell gives this cell an edge with regard to growth. If one cell in this clone mutates again in a way that the chances for growth are increased further, the tumor develops. This in turn offers more possibilities for additional mutations to occur as there are more and more cells in the tumor. Collecting several somatic mutations in one cell takes time. The risk of cancer development, therefore, increases strongly with age.

Mutations causing cancer always occur in genes that control aspects of birth or death of cells. These genes encode proteins involved in cell growth, apoptosis, DNA repair, or control of cell division. There are two kinds of mutations that can result in cancer. The first type activates a given protein or signal transduction cascade no matter whether the signal is there or not. Mutant genes of this kind are called oncogenes and their non-mutated, normal forms are called proto-oncogenes. Oncogenic mutations normally are dominant. Mutations of the ras-gene, for instance, which is involved in numerous signaling cascades, can be found in many human tumors. Tumor viruses can transfer oncogenic versions of these genes and initiate cancer.

A second common type of mutations occurs in "tumor suppressor" genes. The normal function of their protein product is to inhibit growth. When they become non-functional through mutations, the cells start dividing and cancer may develop. Examples are the genes Rb and p53. The Rb-gene encodes a stop signal at the end of the cell division cycle and is involved in practically all cell divisions. If it is missing, the cells continue to divide. The p53-gene checks for mistakes during DNA replication. If mistakes are found, p53 initiates apoptosis, eliminating the defective cells. Mutations in the p53-gene have two serious consequences: first, cells with flawed DNA are not eliminated, thus new mutations accumulate; and second, apoptosis is never initiated and so the defective cells continue to divide. Most human tumors contain p53 mutations.

Many types of cancer are affected by mutations in genes that play a role in specific processes. Colon cancer results from mutations of a wingless signaling pathway, which is normally involved in the continuous regeneration of the colon wall by stem cell divisions. The chance to get this type of cancer is sometimes increased by hereditary disposition, in which one of the six or seven mutations necessary for this kind of cancer is already present in the germ line. Intriguingly, many developmental genes, which have been found in *Drosophila* or *Caenorhabditis*, are homologous to human oncogenes or tumor suppressor-genes. Further research in these model organisms may thus help find other components involved in cancers and may reveal details of the molecular processes that lead to the transformation of individual cells. This is an important prerequisite for developing cancer therapies.

CHAPTER IX

Evolution, Body Plans, and Genomes

ANIMALS DISPLAY AN EXTREME DIVERSITY OF FORM, LIFE STYLE, AND BODY organization. Their ancestors were built simpler than the species existing today. One of the most interesting questions in biology is how these forms evolved over time and which innovations of their body plans helped them adapt to new conditions. The answer to this question is difficult because in most cases, neither the intermediate forms nor the common ancestors exist anymore. Fossils, which are the only clue to the animals of the past, are very rare and tell us almost nothing about the embryonic development or the genes of these animals.

Animals can be classified based on their similarities, and thus their evolutionary relationships, into groups known as taxa. Members of a particular taxon all share a common ancestor. The largest taxa are called phyla, and there are approximately 30 of them today, though there existed probably many more in the past. The five phyla that include the highest number of animal species are the nematodes or round worms, the annelids or ringed worms, the mollusks, the arthropods, and the chordates. These phyla, in turn, are divided into classes, and each class is then subdivided into orders, families, genera, and species. For example, the largest class of the chordates is the vertebrates. The vertebrates themselves are then divided into the five orders of fish, amphibians, reptiles, birds, and mammals. The mammals include the humans, *Homo sapiens*.

As mentioned earlier, the relationship between different species can be recognized more easily in embryos than in adults. This is because during embryonic development an animal's basic construction plan becomes apparent. This body plan is displayed most clearly at a stage when the animal is not yet fully developed and when it is not yet able to feed itself. At this early stage the body is still comparatively simple because the structures necessary for adaptation to a specific life style have not yet developed. Therefore, animals are often grouped as similar based on embryonic structures rather than on adult ones. The vertebrates, for example, are subdivided into amniotes (reptiles, birds, and mammals) and anamniotes (fishes and amphibians) based on whether they possess an amnion—a protective embryonic covering. The amnion, however, is not the only criterion for this classification, as there are also insects that form an amnion. The fact that some vertebrates have an amnion is based on homology, in other words, derivation from a common ancestor, while the presence of an amnion in both vertebrates and insects is based on analogy, namely, similar function that has arisen independently during the evolution of the two phyla.

In many cases, it is difficult to determine whether the animal lacking a particular trait is more basic and its ancestor preceded the one of the animal that has it, or whether the trait had been originally present but was subsequently lost in evolution. Because of these uncertainties, the determination of evolutionary relationships between animals always has to rest on several criteria.

Nowadays, the most reliable criteria are no longer morphological, but molecular characteristics, for example, number of base exchanges in homologous genes. One reliable approach of molecular phylogeny compares DNA sequences of genes that are present in all animals—in particular gene segments that do not code for proteins. In such non-coding DNA segments, changes are assumed to be mainly accidental, without an influence on the phenotype and therefore not subject to selection. Once a mutation in non-coding DNA has occurred, it should therefore be passed on to all descendants. At present, variation in the genes encoding ribosomal RNA is the most commonly used criteria for the taxonomic classification of more distantly related species. For the analysis of the differences between more closely related species, and the variation between individuals within them, mitochondrial DNA sequences are used. Mitochondrial DNA mutates more often than chromosomal DNA, as there are fewer DNA repair mechanisms operating in mitochondria. Therefore, there will be more informative sequence differences occurring in the comparatively short time after the separation of two closely related species.

1. Evolutionary History of Organisms

When genes between different organisms are compared, one striking observation is how similar they are in sequence and often also in molecular function. This is especially surprising in the case of developmental genes that regulate how body plans and organs are formed. These genes are sometimes so similar that they can readily replace each other in two animals that look very different. This suggests that all animals arose from one ancestor that had already a set of developmental pathways and a rather detailed body plan with top and bottom, front and rear, and several organs at specific places. Some such genes can even be traced back further to the single-celled ancestors of animals. The single-celled ancestors in turn are presumed to have their origin in some kind of bacteria-like cell. This is evident from the observation that the elements of the basic metabolic and genetic machinery of a cell are common to organisms with evolutionary paths, which separated billions of years ago, such as humans and bacteria.

Bacteria. Bacteria-like cells were probably the first organisms on Earth. Bacteria are relatively simple cells surrounded by rigid cell walls that determine their shape. Bacteria already feature the basic mechanisms for cell replication such as DNA, RNA, protein synthesis, and ribosomes. They do not yet have a nucleus and their DNA, a ring-shaped molecule, is arranged loosely in the cytoplasm. While there is no true sex in bacteria, a form of genetic exchange does take place between individual cells. In fact, DNA can sometimes be transferred between cells of different bacterial species. This phenomenon is known as horizontal gene transfer. Bacteria do not display a particular diversity of shapes, and although some species form aggregates of cells, there are no truly multicellular species. However, they do possess a remarkably diverse biochemical ability to convert and build materials of all kinds, allowing for adaptation to even the most extreme conditions of life; for instance, there are bacteria that grow best at 110°C (230°F).

Eukaryotes. The first organisms containing a cell nucleus, the eukaryotes, are assumed to have developed more than 2 billion years ago. At first they were only single-celled, just like bacteria, but later several types of multicellular forms evolved. As an evolutionary innovation, eukaryotes have a nucleus that houses the DNA and is surrounded by an envelope. Folding in of the cell membrane allows particles to be moved into the cell. The cell is subdivided into separated compartments formed by membrane-bounded organelles. This restricts the exchange of molecules and provides chemically separated environments within one cell. Another important component of the eukaryotic cell are the mitochondria, which provide energy and most likely originated from bacteria that were enslaved by the ancestral eukaryotes. Eukaryotes have a versatile interior cytoskeleton (rather than the exterior rigid cell wall of bacteria), which stabilizes the cell and at the same time allows for shape changes and motility. The DNA in the eukaryotic nucleus is packed by certain proteins and partitioned into chromosomes to ensure correct distribution into the daughter cells during division. Most components of the eukaryotic cell have been preserved during evolution: the proteins of the cytoskeleton, the enzymes of the DNA replication, and the regulators of cell division; they are still active in today's single-celled eukaryotes, such as yeast, as well as in multicellular eukaryotes, such as plants and animals (see Chapter II, Figures 2 and 3).

The First Animals. Multicellular organisms developed from single-celled eukaryotes. In fact, multicellularity probably evolved several times, in plants and fungi, but only once in animals. The first animals probably looked like a hollow ball and comprised only two layers of cells, one on the outside and one on the inside. The organization of eukaryotic cells into aggregates is facilitated by the loss of rigid cell walls. This allowed for the evolution of cell adhesion and cell communication. Later, animals evolved a branched blind gut and several cell types, among them simple nerve and sensory cells. Such organisms probably looked similar to today's Cnidarians, such as corals or jellyfish.

2. *The Cambrian Explosion*

The first animals can be traced back to more than 600 million years ago. Then, during the Cambrian period, the atmosphere contained levels of oxygen similar to those today, built up by photosynthetic bacteria and algae. This may have allowed animals to grow and spread faster and may have ultimately led to the evolution of a large diversity of forms. During this so-called Cambrian explosion, new organisms with a great variety of body organization and lifestyles emerged within a relatively short period of about 50 million years. As fossils show, representatives of animal phyla still alive today developed then: mollusks, arthropods, worms of all kinds, and chordates. Additionally many other forms that have no similarities to any of the phyla living today existed at that time as well. Their lifestyles remain a mystery.

Bilaterians. The emergence of different forms was facilitated by the evolution of a robust body organization, which allowed for variation in shape and growth. The embryo had become flatter and was composed of three germ layers, now including the mesoderm. The inner cells, derived from the endoderm, formed the through gut. Ectodermal sense organs and a mouth developed at the front, specialized to feed. The mesoderm allowed for formation of muscle and circulatory systems. This prototypic animal, a kind of roundish flatworm, presumably had all essential developmental mechanisms in place. The once radial symmetric animals now could acquire an elongated shape with a clear front and back, top and bottom, and bilateral symmetry. In other words, this primordial animal must have already had all the genes in the most important signaling cascades, such as Delta–Notch and wingless as well as their components, and the gradient system creating the dorsal–ventral axis with Dpp and sog. The Hox-genes, as well as other selector-genes determining the position of sense organs (eyeless) and the heart (tinman), were probably also included in the early repertoire of developmental genes.

Most animal phyla belong to one of two groups of bilaterians: the protostomes or deuterostomes. Their distinguishing characteristic is whether, during gastrulation, the invagination of the gut will eventually become the mouth (protostomes) or the anus (deuterostomes). Vertebrates and echinoderms such as the sea urchin are deuterostomes, while arthropods and mollusks such as mussels and snails are protostomes.

The Inversion of the Axis. In the nineteenth century, French zoologist Etienne Geoffroy Saint-Hillaire (1772–1844) remarked that arthropods are similar to vertebrates with regard to their basic body organization. His example was a lobster. All you had to do to plainly see some fundamental similarities between a lobster and say a human, was to lay it on its back and invert the positions of the anus and mouth. The central nervous system in deuterostomes lies on the dorsal side, but lies on the ventral side in protostomes, such as the lobster. Likewise, deuterostomes have their heart on the ventral side, while protostomes have it on the dorsal side (Figure 54).

Comparing the distribution of certain genes' products in the embryos of vertebrates and arthropods offers many further insights: For example, the succession of the Hox-genes, which pattern the anterior–posterior axis, has been preserved in all animals. Dpp and sog create a gradient that determines the dorsal–ventral axis in arthropods. Vertebrates develop a corresponding dorsal–ventral gradient with BMP (homologous to Dpp) and chordin (homologous to sog). Many molecular feedback mechanisms of this system have been preserved, which suggests that they already operated in the ancestors of protostomes and deuterostomes. But in vertebrates and arthropods, they are arranged in opposing orientation. The BMP signal is ventral in vertebrates and dorsal in arthropods (Dpp) (Figure 54). It is conceivable that a twisting of the body organization would have turned a protostome into a deuterostome. Yet it is also possible that protostomes and deuterostomes developed independently from a common ancestor.

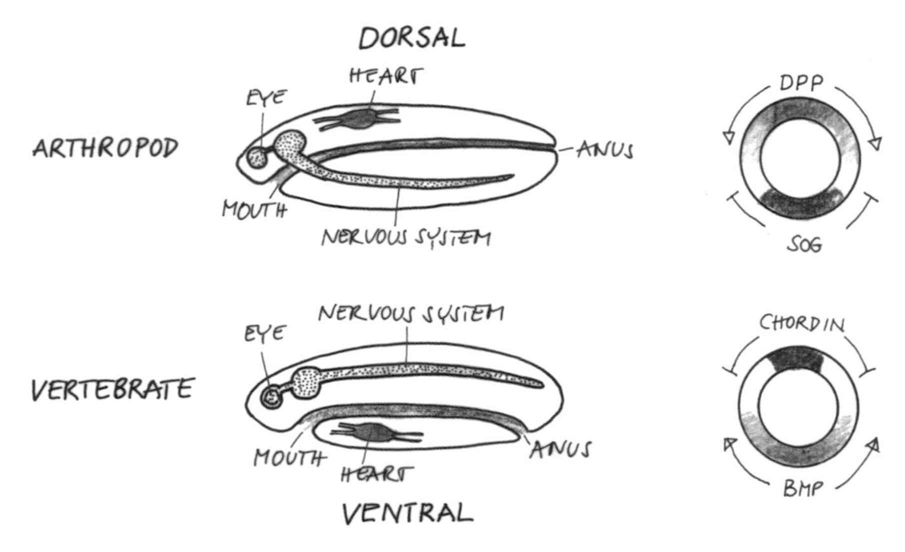

Figure 54. *Inversion of the Body Axis.* When arthropods, such as insects, are compared to vertebrates, many structures seem turned upside down. For example, as illustrated here on the left, the nervous system is on the ventral side in insects and on the dorsal side in vertebrates. Also, the heart beats on the dorsal side in *Drosophila* but on the ventral side in humans. This dorsal-ventral polarity develops during gastrulation by the complex interplay of Dpp and sog in *Drosophila* and BMP and chordin in vertebrates. Sog is expressed ventrally in flies and its homolog chordin dorsally in vertebrates. The inverse is true for Dpp and BMP.

3. New Construction Principles

Segmentation. The variety of animal forms developed due to additional construction principles modifying the basic forms. Segmentation, which divides the body into repetitive units, was one such important innovation. Segmental units, which are initially built based on the same principle, can be modified to form a variety of structures. Indeed, members of many different animal phyla are segmented, but the segments develop in different ways. In annelids and many arthropods, stem cells in a growth zone form precursors of the segments. In *Drosophila*, segmentation results from a simultaneous subdivision of the blastoderm (see Chapter V). Arthropods use the segment-polarity-genes known from *Drosophila* in their segmentation. The genes operating early during *Drosophila* segmentation such as bicoid, hunchback, and knirps have not been conserved in other arthropods, and are likely to be a late innovation of the insects. In vertebrates, segments form in an entirely different way. In this case, a time-delayed pulsating activity translates into spatial waves involving Delta–Notch signaling. In arthropods as well as vertebrates, mechanisms independent from those creating the body segmentation form the head and the frontal segments.

Skeletons. The development of solid support structures facilitates variation in form, and hence adaptation. Cuticle built from chitin, calcium shells, and interior skeletons composed of cartilage and bone offer protection and allow the development of new ways of locomotion. The hard calcium shell of snails and mussels attaches to the body muscles. The arthropods' outer chitin skeleton, with its many flexible joints, allows for sophisticated modes of movement.

The interior skeleton of vertebrates is formed by the mesoderm, which builds cartilage and bones of high strength. Skeletons, by rendering physical support to the body, allow an increase in the size of the animals. Outer skeletons, such as the insect cuticle, provide optimal protection against infection, but they have to be shed when the animal grows. Growth of animals with an inner skeleton, such as vertebrates, occurs continuously, and these animals can grow to very large sizes.

Limbs. Limbs can function as legs, antennae, claws, jaws, arms, or wings, and thus contribute to the variety of forms. Limb development is guided by similar molecular mechanisms as the early stages of body organization. The limbs of arthropods come in an incredible variety of structures, even though they all are formed according to the same construction principle. As part of the head, they are shaped into sensory or defense organs and a rich spectrum of chewing tools. Vertebrates have two pairs of limbs, which can be fins, legs, arms, or wings. Some limbs are homologous among vertebrates, for instance, forelegs, wings, and pectoral fins of fish. Vertebrate limbs develop from buds that arise from the body wall, and their pattern formation involves similar molecules as in arthropods. This may either be due to the common ancestor having had limbs or due to the fact that these signaling molecules may be particularly suited to forming limbs.

Neural Crest. The ancestors of the vertebrates and some other groups of the Cambrium were rather simple chordates lacking a skull, and feeding by filtering prey out of the sea water through their gills. Vertebrates today, by contrast, display a skull protecting a large brain, and a great variety of feeding structures. For this evolutionary innovation, the invention of the neural crest was crucial. The neural crest cells emerge from the dorsal ectoderm near the future nervous system, migrate throughout the body, and can differentiate into many different structures to furnish the body's periphery with many specialized functions. Vertebrate innovations that depend on cells of the neural crest are some bones of the skull, jawbones, and teeth, to facilitate food intake. Neural crest cells also produce the pigmentation of the body, the claws, horns, and beaks.

In general, vertebrates are more complex than arthropods. This is partly because the arthropods' outer skeleton requires shedding of the skin to allow growth. The larger the animal, the more difficult and dangerous molting becomes. This puts a limit to growth. In contrast, the inner skeleton of vertebrates allows for continuous growth, and enormous sizes can be reached. Vertebrate structures such as the neural crest and the placodes further the development of the head and the brain, which reaches an unprecedented complexity in humans.

4. Genomes

The complexity of the body organization of animals is mirrored in the size and number of their genes. The simplest organisms, bacteria, have small genomes with average gene sizes of about 1000 base pairs. The number of genes per genome varies from species to species: the smallest bacterial genomes comprise only about 600 genes, while *Escherichia coli* contains 4300 genes and has a genome size of 4 million base pairs. The human genome, the largest deciphered so far, is about 1000 times the size of *E. coli*'s genome.

Genomes involve dealing with large numbers. The genome of the first multicellular organism sequenced—by the British biologist John Sulston (1942–) and colleagues in 1998—was the genome of the nematode worm *Caenorhabditis elegans*. It contains 100 million base pairs. For deciphering the genome, big overlapping DNA segments were sequenced and computer programs, which recognize regions that can be translated into proteins, determined the position and number of genes.

The *C. elegans* genome has about 19,000 genes. In comparison, the single-celled yeast *Saccharomyces cerevisiae* features about 6000 genes, and *Drosophila* contains about 13,000 genes. It was quite surprising to find that the number of genes in mammals—humans and mice—was not much higher than those found in the fly and the worm: about 32,000. However, mammalian genes are larger than the genes of flies and worms. On average, a human gene contains 27,000 base pairs, and about 10 exons with 100 base pairs each, while introns are about 10 times as long.

Junk-DNA. These comparisons allow for an important conclusion: the growing complexity during evolution does not necessarily correlate with the size of the genome nor does it require a corresponding increase in the number of genes. Although the genes themselves get larger with increasing position along the evolutionary path, the growth in genome size is mainly due to DNA sequences that do not code for proteins and may even have no function at all. About half of the 3 billion base pairs of a mammalian genome consist of such "junk"-DNA. This term mainly refers to short sequences that occur repeatedly in large numbers, so-called repetitive DNA. These sequences may give interesting insights into the evolutionary history of the genome, although the reason for their existence is not clear at all. It is to be expected that organisms with a short generation time lose useless genetic material more quickly. However, it is quite possible that the mammalian genome actually collects junk-DNA, because the junk does not cause disadvantages in the process of selection, and thus can be tolerated.

Some vertebrates obviously preserve large DNA regions that have no apparent function. A comparison between fish illustrates this: the puffer fish Fugu lives with a genome a quarter the size of the zebrafish, although both have about the same number of genes. Comparing the sequence of untranslated DNA regions in two closely related animal species might help to understand their significance. Regions that are similar or the same do most likely have a function, even though this function may not be immediately clear from the DNA sequence.

Gene Families. Comparison of sequences among related genes shows how new genes evolve from existing ones. In this process, three principles are at work: first, mutation of individual bases; second, gene duplication; and third, the new combination of gene parts of different origin. Bacteria feature one additional mechanism: horizontal gene transfer, for example, the exchange of genes between different species. Horizontal gene transfer does not seem to be occurring in eukaryotes.

Most proteins are composed in a modular manner, which means that they contain particular domains of short amino acid sequences that confer specific biochemical properties such as DNA binding or enzymatic activity. About 1200 different protein domains are known, which are used in many different combinations, thus defining many gene families that developed through gene duplication. In homologous genes of different species, often only the sequences coding for these very domains are preserved. The surrounding sequences can be modified greatly, while still retaining the original function of the protein. Of the 1200 domains known so far, about 200 can be found in the proteins of all organisms, from bacteria to humans.

While the proteins of yeast are rather simple, those of multicellular organisms generally contain several domains per gene. These proteins seem to have evolved by combining parts of different genes, such that proteins with completely new functions may have been created. Mammalian proteins are particularly complex. About one third of mammalian genes can be spliced in more than one way such that several proteins are produced which may differ in domain composition and thus function. Therefore, in mammals the number of different proteins is higher than the number of different genes.

Genome Duplication. Gene families most likely arose from duplications of individual genes, which may have been caused by a rare mistake during recombination or replication. But also, duplications of entire genomes are conceivable. The genome of vertebrates, for example, seems to have been duplicated twice in a relatively short period of time during the evolution of vertebrates; with one duplication having been completed already before the appearance of fish. Chordates without a skull, like the lancet, for example, only have one complex of Hox-genes while most vertebrates have four of them. There is a whole range of genes, which appear in four similar copies in the mouse, but only once in *Drosophila*. Frequently, such gene duplicates are preserved only if they can take on different functions through a series of mutations. In fish, there has been an additional genome duplication, but only about 20% of the duplicated genes were preserved.

Many genes and their properties had been known even before systematic international projects managed to sequence entire genomes. Nevertheless, such projects are very important because they allow for elucidating which genes are present in which organism and, more importantly, which ones are not. The resulting data have made it easier to find homologous genes, which—due to their divergence in sequence—can only be identified through computer-supported search programs.

Two mammalian genomes have been deciphered: that of the mouse and that of humans. As expected from their close evolutionary relationship—the last common ancestor lived about 100 million years ago—their genomes are relatively similar. At least 99% of the mouse genes have a counterpart in human genes and 78% of the amino acids in the respective proteins are the same. These numbers do not say much though, as the degree of similarity varies considerably from gene to gene, and because functional variations may depend in complex ways on the primary sequence of the protein. This comparison also does not take control regions into account. But important differences between humans and mice are revealed in the number of genes influencing certain protein families. For example, mice have many more genes involved in sensing smell than do humans.

Sequence comparisons among eukaryotes show that more than 60% of *Drosophila* genes and almost 50% of worm and yeast genes are homologous to mammalian genes. This does not mean that these genes in all cases fulfill a similar function, since many mutations have taken place to adapt proteins to new processes more relevant to the respective organism. But it does mean that these genes share a common origin. These numbers stress the conservative nature of evolution, which nevertheless in its long course has brought forth an amazing variety in form, function, and lifestyle of organisms.

5. The Evolution of Humans

Molecular biology can also be used to investigate the origin and evolution of humans. An analysis of the differences in the DNA between people existing today is quite revealing. In a few cases, a comparison of the DNA extracted from human fossils with that of humans living today has been successful as well. However, most of what we know about ancient human evolution rests on fossil evidence, which is naturally shaky due to the scarcity of human fossils.

Homo Sapiens. Fossils have shown that the first primates existed about 60 million years ago and later evolved into hominids in Africa. The oldest fossils suggesting an upright gait are about 4 million years old. It could not be clarified, however, when and where the modern human, *Homo sapiens*, arose, because fossils of the genus *Homo* have been found on several continents.

The analysis of differences between DNA in humans of different ethnic groups living today has solved the issue. The most extensive data stem from the comparison of mitochondrial DNA. Mitochondria are passed on exclusively by egg cells, and hence, strictly speaking only reveal the relationships among female ancestors. Male relationships are studied by examining certain regions on the Y-chromosome, which are less variable and also do not undergo the process of recombination.

There are variations in about 3 of 100 base pairs within a mitochondrial DNA sequence. It is possible to reconstruct a family tree from the common variations in humans living today. The largest degree of variation was found among Africans, while only subsets of this variation were found in non-Africans. This indicates that the origin of all modern humans is Africa—it is from there they emigrated and spread all over the world. Considering the relatively small number of variations, the species *Homo sapiens* can be traced back to a surprisingly small founder population of probably no more than 10,000 individuals. The other striking finding stemming from such comparisons is that there is more variation between people within a given population than between different populations. This is surprising because, superficially, people from different parts of the world can look rather different in terms of body size or skin color. However, their overall difference in the genes is tiny compared to the differences between individuals within one ethnic group. Thus, there is no real genetic basis for the idea of human races.

When did modern humans evolve? One estimate is based on the differences between humans and chimpanzees, the human's closest living relative. The DNA variation between humans and chimpanzees is roughly estimated to be about 25 times greater than between any two humans. These differences arose after the split of the lineages leading to humans and chimps, respectively. According to age determination of fossils, the common ancestor of humans and chimpanzees lived about 5 million years ago. Thus, *Homo sapiens* emerged about 200,000 years ago (5 million divided by 25 equals 200,000). Considering also information based on other methods, we end up with an average time span of 150,000 years ago.

Fossil discoveries in central Europe show that modern humans lived there about 30,000 years ago. These humans already produced art, as is evident from many impressive cave paintings in southwestern Europe. At the same time, members of a different human species, known as Neanderthals, named after the Neanderthal in Germany where their fossils were first found, populated Europe as well. Comparing Neanderthal DNA sequence extracted from fossils with the sequence of modern humans showed 10 times more deviations than within individual modern humans. This means that the Neanderthals most likely were not a direct ancestor of modern humans. The last common ancestor of modern humans and Neanderthals is assumed to have lived about 500,000 years ago (Figure 55).

Primates. The closest living relatives of humans are the great apes: chimpanzee, bonobo, orangutan, and gorilla. Molecular data show that we are most closely related to the chimpanzee and least closely related to the orangutan. Chimpanzee and human genomic DNA sequences differ only in one of 100 base pairs. By comparison, two non-related human beings differ in one of 1000 base pairs.

Very recently, the genome of the chimpanzee was completely deciphered. Although humans differ from chimps by only about 1% of the bases, this still amounts to tens of millions of individual bases because these genomes are so

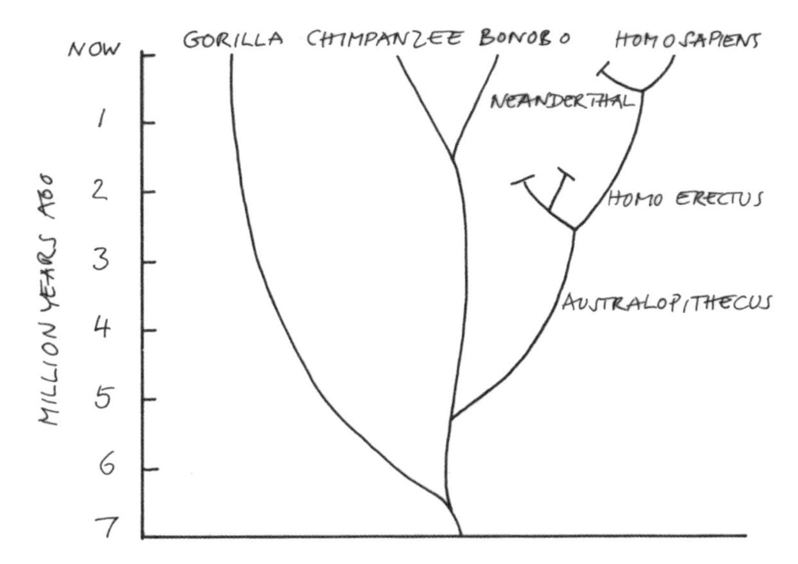

Figure 55. *Family Tree of Hominids.* The oldest common ancestor of primates and humans lived about seven million years ago. Modern man developed probably less than 200,000 years ago. All other hominids, such as the Neanderthals and Australopithecus, have died out.

big. About one-third of the proteins present in chimps and humans are identical and most of the rest only differ by one or two amino acids. This means that the differences at the molecular level are very subtle. Why then are humans and chimps so different? Comparing the genomes of these species will help to answer this question by identifying regions, genes, or regulatory DNA that have been under intense selective pressure and that differ relatively greatly from chimps but are very similar between different human groups.

It is interesting that the DNA variation within one species of ape living today is three times greater than those among humans, even though the population of living chimpanzees, about 50,000, is much smaller than that of the 6 billion humans. This suggests that the ape species are older and that they never reproduced rapidly. Human populations, however, expanded quickly and, as previously stated, probably originated from a very small founder population 150,000 years ago. Therefore, all people living today are very closely related genetically. Until, and even after, the emergence of modern humans, the population was relatively small. But with the end of the last ice age, and with the subsequent invention of farming 11,000 years ago in the Fertile Crescent, now known as Iraq, the human population blossomed. Archeological findings support these theories.

CHAPTER X

Current Topics

IN RECENT YEARS, GENES AND EMBRYOS HAVE BEEN THE SUBJECT OF INTENSE public discussion. This is mainly due to the fact that achievements and scientific discoveries in the fields of embryology and genetics not only increase our knowledge, but also open up new possibilities to influence human life in a principally novel way. In addition, these achievements give rise to speculations and scenarios, which would indeed change our world substantially should they ever be realized. Although several medical applications of new technologies, in particular gene technology, are now widely accepted, there is a widespread fear of the dangers of unpredictable consequences of such technologies. At the center of these debates is the issue of the extent to which human embryos should be manipulated in vitro and whether or not to interfere with their genetic constitution. Between different countries, the regulations relating to this issue are diverse, ranging from very restrictive, for instance in Germany and Ireland, to quite permissive in the United Kingdom and Sweden. It is apparent that the debates will continue as long as it is not attempted to reach a consensus and formulate reasonable rules for dealing with human embryos that apply to the research in all countries.

What exactly is the issue? Since 1978 it is possible to fertilize human eggs outside the female body and to cultivate the embryos in vitro for a short while, before they are transferred back into the uterus to achieve pregnancy. The current procedure for this in vitro fertilization (IVF) yields surplus embryos that are not transferred back. This opens up the possibility to use these embryos for medical research rather than discarding them. For instance, embryonic stem (ES) cell cultures could be obtained that could be used to develop therapies for several severe diseases. By genetic screening of such early embryos, congenital diseases could be avoided, even eradicated. However, according to the generally accepted moral conviction of our culture, every human being possesses dignity and, therefore, must not be used solely for the benefit of others. Of course, both opponents and supporters of embryo research share this moral position. The conflict, therefore, does not concern the acceptance of human dignity and protection, but rather the moral status of the early embryo.

From which point on is a human embryo a human being? Are the very early embryos human beings, which have to be protected just like growing embryos during pregnancy or newborns? Or is their moral status different, and gradually changing until birth? The view of such a graded increase in status in many ways does reflect our natural feeling, which is manifested by our customs of birth control and laws of abortion. And this difference in status is the bone of contention. The German law of 1990, for instance, defines the beginning of a human being at fertilization, while others consider the actual time of implantation into the female organism as crucial. In the United Kingdom, research on embryos in vitro is allowed until the 14th day. A human being becomes a legal person only by birth.

Although the criteria used for the definitions rely on biological events, they are not a scientific but a moral issue. Dignity, right to life, and protection are not biological, but moral categories. Therefore, these issues should be decided not by scientists but by our society as a whole through our political representatives. The big differences between nations, even of very similar cultural backgrounds, indicate that there is no single right solution. However, in order to guarantee efficient research, we need clear regulations that are widely respected. The legal definition of the beginning of a human being should be reasonable, plausible, and consistent. It is at this point where scientific knowledge and judgment may help by describing grades and steps of embryonic development. It is also the obligation of scientists to reveal the potential applications and consequences of embryonic research. As there are conflicting moral issues such as the right to life, on the one hand, and the pain and suffering caused by yet untreatable diseases on the other, such regulations demand great care and foresight and will have long lasting consequences to our societies.

Using early human embryos for medical therapy is only part of the debate. Additional issues include the ability to interfere with the processes of reproduction and diagnostic procedures, such as the selection, or even the genetic manipulation of children with desired attributes. A particularly controversial issue is that of cloning. It is remarkable that in this debate, or rather in the representation of this debate by the media, there is little or no distinction between what is real, what is plausible, and what is utterly utopian. Sometimes the impression is conveyed that science could not only accomplish anything we desire, but would actually test the limits of the possible without any ethical considerations. This attitude is not new. There is a long tradition on both sides: blind belief in scientific progress and its flip side, the utter distrust in science.

1. Utopias

Utopias of man creating man have existed since Antiquity, handed down through the generations by way of myths and religions. But even though most people would acknowledge that the creation of woman from Adam's rib or the creation of Pallas Athena from Zeus' head are meant purely as symbols, things look different if such ideas are supported by contemporary biological theories. A good example from the Middle Ages is the recipe of the creation of the homunculus (Latin for little human) by Paracelsus of Hohenheim, a scholar of alchemy and various other sciences in the sixteenth century. The story of the homunculus is based on the popular idea of preformation, which prevailed for a long time (see Chapter I). This idea implies that the sperm already harbors a completely formed human being, a homunculus, which unfolds in the mother's body in the way a plant's seed would develop in the earth. Paracelsus' 1537 recipe replaces the mother's organism with an artificial medium. The sperm is incubated in a concoction of horse dung, urine, and other ingredients, and kept warm inside a pumpkin. According to Paracelsus, a small human being would appear within 40 days, provided that the process was undertaken in secrecy. The second part of Johann Wolfgang von Goethe's (1749–1832) classic play *Faust* picks up on this story in a remarkable scene. It is not Faust, the dropout and skeptic, who kindles the fire, but Wagner, the zealous and progress-oriented student:

> *Mephisto: What, pray?*
> *Wagner (softly): A human being in the making.*
> *Mephisto: A human being? Have you a loving pair*
> *Locked in your chimney, in their tender passion?*
> *Wagner: Now God forbid. That old style we declare*
> *A poor begetting in a foolish fashion...*
> *What if the beasts still find it their delight,*
> *In future man, as fits his lofty mind,*
> *Must have a source more noble and refined.*

However, in the play, the experiment fails because homunculus, a precocious but underdeveloped creature, albeit one adored by Wagner, cannot climb out of the test tube. He appears on Earth only "half-baked." Aided by

Mephisto, homunculus looks for advice from the wise naturalists of the ancient Greeks, and it is Thales of Milet who finally throws him into the sea with the words:

> *Submit to a request to winning,*
> *And start to be at the beginning.*
> *Accept swift working of the plan:*
> *Then, following eternal norms,*
> *You move through multitudinous forms,*
> *To reach at last the state of man. (Faust II, Act 2)*

This text harbors several truths: the truth of evolution, that is, the development of complex forms from simpler ones over a long period of time; the truth of defined rules according to which life develops; and finally, the vanity reflected by the scientist's (Wagner's) attempts to fathom these norms for his own purposes, his own creations.

While the homunculus story belongs to the realm of fantasy, British writer Aldous Huxley's (1894–1963) utopia of human creation as described in his novel *Brave New World* (1932) is often considered to be rather realistic—at least if it is not possible right now, it is seen to be possible in the near future. Huxley imagines a procedure by which embryos can be made to form buds, such that from one embryo several identical copies can develop. These cloned embryos are then to be raised in bottles that serve as a kind of artificial uterus. The realism of his story is buttressed by Huxley's detailed description of the physical and technical difficulties of the uterus machine. Most striking is perhaps his conjecture that the conditions under which the clones mature can be manipulated to program the desired attributes of these parentless beings. In Huxley's time, the conditions of human and mammalian development were known only in broad strokes and the nature and biochemical function of genes were not known at all. Yet, it is even more remarkable that there are people who take Huxley's ideas as present or future truth rather than as the fiction that it is.

2. *Cloning*

While our understanding of the general processes involved in human development has been aided by research on model organisms, such as the mouse, procedures discussed for application in humans are based on research on domestic animals. Artificial fertilization as well as genetic diagnosis of single embryonic cells has been developed and studied in cattle. The original idea was to use genetic diagnosis to predetermine the sex of a calf in order to produce predominantly female cattle for milk or male cattle for meat, depending on the breed. But the procedure is now rarely used because it is too complicated.

For artificial fertilization, hormone treatment of the cow stimulates the production of supernumerary eggs. Then the eggs are removed from the ovary and placed in a culture dish where they are either mixed with the sperm, or where the sperm is injected directly into the eggs. After a few days, when the fertilized egg has divided several times, the embryo is reimplanted into the mother cow. For genetic diagnosis, prior to implantation, one or two cells are removed from the embryo and its chromosomes are analyzed. The removal of one or even two cells at this stage does not harm the embryo, and it will develop normally after implantation.

Cloning Animals. The first experiments in cloning by nuclear transplantation were conducted in the 1960s on amphibia (see Chapter II). Researchers were interested in determining whether all body cells maintain all genes that are necessary to create a healthy animal. The first mammal to be cloned was a sheep called Dolly. Breeders' interest in cloning lies in multiplying genetically identical animals, which have proved to have particular desired properties. This is also the basis for the wide use of cloning in crops. However, plants are not cloned by cell nucleus transfer but rather by taking layers and cuttings for plant reproduction, a quite natural event propagating genetically identical plants.

During the procedure of animal cloning, the nucleus of an egg cell is removed and replaced by a nucleus taken from a body cell of a chosen animal. In rare cases, a blastocyst will develop, and even rarer still this blastocyst will give rise to a healthy animal. Such a cloned animal carries the same genotype as the donor animal from which the nucleus originated. Even though cloning has been accomplished in several animals—among them cows, sheep, and mice—the success rate is extremely low. In most cases, the clone's development is sooner or later interrupted, resulting in frequent miscarriages and stillbirths.

While these attempts at cloning animals have provided a satisfactory answer to the question of whether the set of genes remains complete in body cells, cloning does not work efficiently for the actual reproduction and breeding of animals. There are several reasons for this. First, the body cells from which the nuclei are taken may have accumulated too many mutations. During normal development, special cells from the germ line produce the offspring. These cells are well protected, leading to fewer mutations than body cells. Second, the developmental potential of body cells is restricted because their genes are wrapped in special proteins and partially modified. These restrictions would have to be reversed completely during contact with the cytoplasm of the egg. This reprogramming apparently takes place only rarely. A probable third reason is that in the egg the chromosomes are not always distributed in an orderly fashion after the nuclear transfer. Whatever the reasons may be, the fact is that cloning by nuclear transfer is successful only in very rare cases. Presently, it is unpredictable how the procedure could be made more efficient and safe.

Cloning of Humans? Cloning a human means creating a person with exactly the same genotype as an already existing person—a belated twin, as it were. To construct such an embryo, the nucleus of a body cell would have to be transferred to an egg cell from which its own nucleus had been removed. But, as previously stated, cloning animals with this procedure only rarely produces healthy animals. In Dolly's case, more than 200 eggs had to be treated before success was achieved. This rate is much too low to justify even the slightest attempt at cloning a human being. With humans, vastly different safety requirements are appropriate than are those for domestic animals. While it is a moot point to debate the ethical implications of a procedure that, although theoretically possible, in reality cannot be performed successfully—at least not for the time being, human imagination has still stirred up extensive ethical discussions on the issue of cloning. Biologically speaking—apart from the extremely low success rate and frequent mishaps predicted to go with this procedure—the fact that a cloned child would have no parents creates a high level of discomfort. In addition, the motivation for a desire to double oneself or somebody else does not bode well for the child's welfare. Therefore, on ethical grounds alone, attempts at cloning of human beings (reproductive cloning) have been rejected by scientists and researchers all over the world. In many countries it has even been rendered illegal.

3. Human In Vitro Embryos

Basic research aimed at understanding the general principles of life uses model organisms such as the frog, chicken, fish, and mouse. Regarding human embryos before implantation, the debate centers on problems of medical research. This research concerns most notably reproductive medicine and cell replacement therapies for degenerative diseases.

In Vitro Fertilization. The procedure to fertilize human eggs in vitro and to cultivate them before implantation was first performed in England. In 1990, legislation was passed there allowing research on preimplantation stage embryos, as long as it serves the improvement of IVF procedures to treat infertility. Convincingly, it was argued that only tried and useful therapies should ever be offered to patients. In contrast, the German embryo protection law from the same year does not permit any research on embryos, even though IVF is widely practiced in this country as well. In the United States, no particular restrictions exist.

Infertility has various causes not all of which can be overcome by IVF. In about half the cases, the sperm are either barely or not at all able to fertilize an egg. In these cases, fertilization can be achieved by injecting sperm into the egg. The German procedure prescribes that all of the usually 8–12 eggs harvested after hormone treatment of the woman be fertilized in vitro. Three are then implanted, while the remaining eggs are frozen before the two "pronuclei" (see Figure 51) have combined at about 20 hours after fertilization. According to the German law, this is when protection of human life must begin; therefore, before this point these embryos are not yet human beings in a legal sense. In England, all eggs are cultured up until the morula stage reached at about 4 days. Of these embryos, the one or two that have developed best are implanted, and the others are frozen. The frozen eggs can either be implanted in a later attempt, or, if permitted by the parents, used for medical research. In the United States there are no strict rules, and it is up to the fertilization clinics to decide how they help women to get pregnant.

Preimplantation Genetic Diagnosis (PGD). Research on IVF intends to improve its safety and efficiency. Preimplantation genetic diagnosis (PGD) has been developed, in which at an early stage one or two cells out of eight cells total are removed from the embryo and tested for their genetic constitution. As mentioned, this procedure does not harm the embryo. PGD may serve two purposes.

First, one can test whether the embryo contains the normal number and types of chromosomes, as abnormal chromosomal content known as aneuploidy resulting from failures during meiosis (see Chapter VI, Figure 39 and Chapter VIII) is curiously frequent in humans. It is in fact one of the major reasons for miscarriages and low efficiency of IVF, especially with older women. By implanting only embryos with normal chromosomal content, PGD can increase the rate of successful pregnancies. It can further help to reduce the high incidence of the very problematic twin or triplet births. These occur because usually more than one embryo is implanted—in Germany three embryos as a rule, in the United States even more—in the hope that at least one will survive.

Second, in rare cases in which both parents are carriers for the same recessive congenital disease, according to Mendel's laws, the embryo has a one in four chance of being affected. In such cases, the diagnosis will help to ensure that only embryos that are healthy with regard to the particular disease gene are implanted. These are rather rare cases, which often are only recognized because the parents already have one diseased child. Potential parents with a disease in the family would first undergo a gene test and in the embryos only the relevant genetic constitution would be examined. In countries with restrictive embryo protection laws, PGD is not permitted. However, through prenatal diagnosis (PND) diseased embryos can be genetically diagnosed at a later stage. In these cases the pregnancy is usually artificially terminated. PGD could thus help to prevent the abortion of fetuses in rather late stages of development. However, for the detection of diseased embryos by PGD the parents have to undergo the procedure of IVF and several embryos have to be tested. Although only one or two cells from each embryo are analyzed, the procedure is quite safe.

4. Designer Babies

One reservation against preimplantation diagnosis is the fear that the spectrum of applications may be widened, and also genetic dispositions other than those immediately concerning the health of the baby may be tested for and used for selection. In this vision of a "designer baby" a different set of genes would be selected for, not those that when defective cause disease, but those that in a more subtle way are thought to determine favored and positive properties of humans, such as intelligence or beauty. However, very little if anything is known so far about such genes and there are hardly any possibilities to recognize them. Experiments with mice are only helpful with genes involving basic biochemical functions that are similar between mice and humans. After all, mice do not possess the cognitive abilities or physical features considered beautiful and desirable in a human being. It is not known a priori which properties a human being will have even if its DNA is analyzed, and even less is known about how it would develop if one or more of its genes would be present in a different constitution.

At this point several limitations become apparent that set boundaries to our visions: (1) Only those gene variants that are transferred into the embryo by the parents can be a basis for selection. (2) One can only diagnose the gene variant, not the property. And the relationship between gene and property is very complex. Furthermore, as every gene is present in duplicate, a gene test would require the analysis of the gene variants, not the presence or absence of a given gene. (3) Many properties are caused by the interplay of several genes. In most cases, these are located on different chromosomes, and accordingly distributed independently into the germ cells. For pure statistical reasons the desired combinations will be very rare. (4) Even in the case of twins with complete genetic identity, the properties of one twin do not permit a safe prediction about those of the other. This means that external influences are also decisive.

For similar reasons ideas about improving human beings by introducing extra genes into their germ line belong truly in the realm of science fiction. Granted though, continued research will increase our knowledge. Still, presently it cannot be foreseen how and when the exact function of human genes and the correlation with desirable and undesirable properties of our children can be elucidated. As it stands now, the idea of "designer babies" is quite unrealistic.

5. *Gene Therapy*

Gene therapy in the germ line means that a genetically defective individual would receive the correct gene so that the individual, as well as his or her offspring, would be free of the disease forever. This sounds attractive. However, presently there is no procedure known allowing the introduction of exactly one copy of a particular gene such that all cells receive this gene and no unwanted side effects occur. But it apparently is possible to create transgenic mice, fish, or flies. So why not humans? The reason is that a gene transfer is a very rare event and will be successful only in a few of the progeny of the treated individual (see Chapter III, Figure 18). This means that experiments with transgenic animals require several generations and many individuals, and are accompanied by many failures. The requirements of safety and efficiency for humans are totally different from those for experimental animals and it is impossible to treat one individual egg, animal, or human, with practically 100% certainty of success. Consequently, what "works" in the case of animals is by no means practicable in humans. In addition, the treatment would have to be done at the one-cell stage at which it cannot be predicted whether the egg is defective or normal. And if the parents are healthy, some of the siblings should be healthy as well. These may be selected by PGD, thus rendering a risky procedure unnecessary.

In several genetically caused diseases it is principally possible to achieve a cure by introducing into patients cells that carry the normal copy of the gene. This procedure is called somatic gene therapy. With such diseases, it suffices in principle if a fraction of the body cells are normal and can shed a protein, which rescues also the mutant cells. The development of such therapies started very early after the first disease genes had been isolated. Although there are no ethical constraints, and the procedures have been intensely worked on, there are as yet only very few successful cases of somatic gene therapy. Furthermore, the present protocols for gene therapy are not completely safe, as there is a risk to develop tumors caused by incorrect integration of the gene into the genome. Somatic gene therapy serves as an example of the difficulty of predicting the success of scientific research. Twenty years ago, at the time when research was started, it seemed straightforward and only a matter of years until an effective therapy was developed. At present, it is far from clear that gene therapy will ever work on a large scale. This should have a sobering effect on too optimistic promises of eventual cures from modern biomedical research.

6. *Human Embryonic Stem Cells*

Embryonic stem (ES) cells from mice have been described in Chapter VIII. These cells can be propagated in vitro without losing their pluripotency; moreover, extra genes can be introduced into the genome of embryonic stem cells using homologous recombination. Cultures of human embryonic stem cells have first been described in 1998. They are not easy to establish and because of ethical problems research on them has not yet progressed far. It is not even possible to be certain whether human embryonic stem cells have a similar potential as those of mice. This is because the test for pluripotency, the creation of chimeras, cannot be performed in humans as it can be in mice. Possible medical applications include treatment of diseases in which certain cell types degenerate and cannot be regenerated by the body, for example, diabetes Type 1, Parkinson's disease, and multiple sclerosis. For these diseases there is so far no sustainable therapy. Research on mouse embryonic stem cells reveals how these cells may be useful in medical therapies. Recently, researchers succeeded in producing special neural cell types from embryonic stem cells of mice that were able to cure symptoms when transplanted into the brain of animals with a condition resembling Parkinson's disease. Preliminary, promising results with embryonic stem cells carrying an extra copy of the insulin-gene were also obtained in the case of diabetes.

There are also other stem cells in the body, so called somatic or adult stem cells. They multiply typically by asymmetric divisions, such that one daughter cell remains a stem cell while the other develops, often after a few divisions, into a differentiated cell replacing cells in the body when needed (see Chapter VI, Figure 40). In contrast to embryonic stem cells, their developmental potential is already restricted. Furthermore, they do not readily divide in culture and are not easily channeled into a different fate. Experiments comparable to those mentioned for embryonic stem cells failed so far, although research on adult stem cells is lavishly supported. It seems likely that for some severe diseases a therapy might first be developed using embryonic stem cells before using other stem cell types.

One problem with potential therapies involving embryonic stem cells is their immune compatibility. As is true for tissue and organ transplants, transplanted cells also have to be compatible with the recipient's immune system in order to not cause fatal rejection reactions. The problem of immune compatibility of embryonic stem cells with the body cells of a patient can in theory be prevented by a procedure called therapeutic cloning, or somatic cell nuclear transfer (SCNT). This involves the construction of blastocysts for medical purposes by nuclear transfer into an enucleated egg cell. The nucleus would be taken from a somatic cell of the respective patient, while the egg cells would have to be donated by a woman. The aim of this procedure is to establish embryonic stem cell cultures that are genetically identical to the cells of the patient and which would not be rejected by an immune reaction. Recently, such nuclear transfer blastocysts have been obtained and embryonic stem cells successfully established from them by South Korean scientists, although with a low success rate. An alternative approach is to select a cell line that is tolerated by the host's immune system from a collection of embryonic stem cells. Such embryonic stem cell banks are being established in England and South Korea. This is similar to the procedure of selecting compatible organ donors in large databases.

As mentioned, the regulations of research on human embryonic stem cells differ widely between countries. In the United Kingdom and Sweden, although reproductive cloning is banned, research on SCNT, or therapeutic cloning, is permitted because the transient blastocyst stage is not yet considered an embryo which has to be protected, but rather is regarded as a clonal extension of the donor. In the United States, researchers funded by public organizations are only permitted to use human embryonic stem cell cultures established before August 1, 2001, while privately funded research is allowed to take place without restrictions. A respective regulation in Germany allows the import of embryonic stem cells that have been established in other countries before January 1,

2002, but the establishment of human embryonic stem cell lines is prohibited. This is an unreasonable compromise, which reflects a deep distrust by politicians toward science and which will effectively prevent scientists from participating in this research in a competitive manner. On the other hand, it is evident that therapies, should they be developed for example in the United States with private funding, or in the United Kingdom, Sweden, or South Korea, cannot be kept from patients living in countries with a restrictive legislation. It seems morally questionable to profit from research performed in other countries, which is not permitted in one's own country, as is already happening in the case of IVF. In light of this, it should be highly desirable to participate in such research now, even before successes have been achieved, not least in order to share the risk of this very promising, but difficult and expensive research between the various countries.

7. *The Moral Status of the Embryo*

The crucial issue causing all of the aforementioned discrepancies in legal regulations regards the question to which degree and from which stage human embryos have to be protected. When does a human embryo have to be protected against destruction and use for other's purposes? As mentioned, it is not the task of the scientists to decide this question, but rather that of society as a whole, and the differences in political opinions reflect by no means a discrepancy of the opinions of scientists in different countries.

Nevertheless, in defining the moral status of the embryo, the classical arguments of ethicists and philosophers are often based on pieces of biological evidence, which sometimes prove to be highly debatable if inspected more closely from the view of modern biology. For instance, a widely accepted dogma argues that human life is a continuous process starting with fertilization that does not display any sharp transitions and during which nothing substantial is being added that would justify a change in status. Another argument states that the zygote with its complete genetic constitution would also hold the complete potential to develop into a human being. Now it is quite clear that a chicken or frog embryo starting with fertilization has the potential, even without motherly protection, to develop continuously until hatching. In the case of mammals and thus also humans, however, the embryo has to implant into the uterus of the mother to be able to develop further. The zygote alone has only the potential to form a blastocyst that has to then hatch from the egg case to implant into the uterus, and begin the next stage of development. Biologically speaking, this is a marked transition and there is almost nothing more discontinuous than such a process in which the embryo is placing itself in direct and immediate cell contact with another individual. In the fertilized egg, the genetic program is complete. But for its realization, the intensive interaction—the symbiosis with a second organism, the mother—is required. This is indispensable and cannot be provided by surrogates. Only at birth, the growing human being has become a separated, independent organism that breathes and now has its own independent metabolism. Certainly the human being born still has much need of attention and protection, but it is now fed from the outside and therefore in case of necessity can survive without the mother. There is no debate that at this stage it is a human being with all rights.

As already mentioned, it is remarkable that the issue of embryo protection is treated in such a different manner in different countries. This reflects the difficulty in compromising between extreme positions. It would be most desirable if one could agree on rules guiding embryonic research that are based on plausible and reasonable grounds. Science is international, and progress in the long run depends on equal and just conditions for scientific research. Such rules certainly must serve to prevent misuse, but they also should not unduly inhibit medical research that is guided by the ethical principle to help and cure existing human beings. In addition to the scientific quality of the intended research, it should be required that sufficient animal experiments have been carried out in order to guarantee a reasonable rate of success to make a procedure practicable in humans. Also, the laws should prevent those embryos manipulated in vitro, for example chimeras with embryonic stem cells, or those that have been constructed by nuclear transfer, from being implanted into a female organism to start pregnancy. Such a regulation would also prohibit reproductive cloning of humans. The most important factor, however, is to proceed with care and ensure that possible contributions of medical research to reduce pain and suffering are not prohibited for fear of misuse.

Appendix

TIME TABLE

Ca. 323 BC—Description of development and heredity: Aristotle, Lykeion, Athens, Greece (see page 2)

1735—Categorization of animal and plant species based on similarities: Carl Linnaeus, University of Uppsala, Sweden (see page 2)

1784—Discovery of the human dental bone: Johann Wolfgang von Goethe, Weimar, Germany (see page 5)

1827—Human egg: Karl Ernst von Baer, University of Königsberg, Prussia (today: Russia) (see pages 3 and 11)

1852–1855—Cells originate exclusively through cell division, *Omnis cellula e cellula*: Robert Remak, University of Berlin, Rudolf Virchow, Charité, Berlin, Germany (see page 11)

1859—*The Evolution of Species by Natural Selection*: Charles Darwin, Down House, Kent, England (see page 3)

1866—*Experiments on plant hybrids*: Gregor Mendel, Augustinian Cloister, Brünn, Bohemia (see page 6)

1869—Isolation of nucleic acid: Friedrich Miescher, University of Tübingen, Germany (see page 31)

1885—The Germ line Theory: August von Weismann, University of Freiburg, Germany (see page 19)

1888—Fertilization by the fusion of egg and sperm, Oskar von Hertwig, University of Jena, Germany (see page 14)

1900—Rediscovery of Mendel's laws (see page 9)

1902—Individuality of chromosomes: Theodor Boveri, University of Würzburg, Germany (see page 16)

1903—The chromosomal theory of heredity: Walter Sutton, University of New York, USA (see page 16)

1910—The polarity of the egg of the sea urchin and roundworm (*Ascaris*): Theodor Boveri, University of Würzburg, Germany (see page 16)

1911—The *Drosophila* white mutant, the X-chromosome, and genetic linkage: Thomas Hunt Morgan, Columbia University, New York, USA (see page 26)

GLOSSARY

Activator	transcription factor that stimulates transcription
Allantois	extraembryonic sac serving the excretion and exchange of air in the chick and mammalian embryo
Allele	state or condition of a gene
Amnion	extraembryonic coverings of the chick and mammalian embryo
Aneuploidy	too few or too many chromosomes
Anterior	at the front
Apoptosis	programmed cell death
Arthropods	group of animals including insects, crustaceans, and spiders
Axon	extensions of nerve cells; nerves are bundles of axons
Bacteria	simplest living organisms, do not contain a nucleus
Blastocyst	stage of mammalian embryonic development; hollow ball composed of trophectoderm and inner cell mass
Blastoderm	stage of *Drosophila* embryonic development during which all cells are still undifferentiated
Blastoporus	ring-shaped region in frog and fish embryo at which cells migrate inward during gastrulation
Blastula	stage of embryonic development in frogs and fish during which all cells are still undifferentiated
Centrosome	cell organelle; starting point of the microtubules that form the spindle during cell division
Chimera	organism composed of cells of different individuals
Chorion	outer shell covering embryos
Chromatid	after doubling, every chromosome forms two chromatids
Chromosome	threadlike structures in the nucleus that carry the genes of an organism

Cleavage	cell divisions after fertilization that divide the embryo into many small cells
Clone	group of genetically identical cells or organisms
Code	translation rules
Cortex	net of microfilaments underneath the cell membrane
Cytoplasm	cell sap, viscous fluid content of a cell outside of the nucleus
Cytoskeleton	threadlike chain molecules running through cells such as microfilaments and microtubules
Diploid	each of the different chromosomes occurs twice
DNA	deoxyribonucleic acid, the double-stranded chain molecule forming the genes
Domains	short stretches within a protein that have certain functions; present in many different proteins
Dorsal	on the back side; there is also a *Drosophila* gene called dorsal
Ectoderm	outer cell layer of the embryo that will later form the skin and nervous system
Embryonic stem cells	cells from the inner cell mass of a blastocyst that can be multiplied in a culture medium and differentiate into all kinds of cell types
Endoderm	inner cell layer of the embryo that will later form the digestive tract
Enhancer	DNA region of a gene that binds proteins that stimulate gene activity
Epithelium	dense two-dimensional cell layer
Eukaryotes	organisms built from cells which contain a nucleus surrounded by a membrane
Exon	DNA region of a gene that is translated into protein
Extracellular matrix	network of glycoproteins that fills the spaces between cells in tissue
Fertilization	fusion of egg and sperm
Gametes	germ cells, eggs, or sperm
Gastrulation	formation of the three cell layers of the embryo; the endoderm and mesoderm move inside the embryo and are covered by the ectoderm
Gene	hereditary unit consisting of DNA that codes for a protein; some genes code for RNA

Gene activity	a gene is active if it is read and translated into a protein
Genome	the totality of all genes of an organism
Genotype	condition of the alleles for any given gene
Gonads	sex organs: female ovaries, male testicles
Gradient	gradual change of concentration of any given material
Haploid	each of the different chromosomes occurs once
Heterozygote	the two copies of a given gene are different
Homeodomain	a protein domain appearing in many transcription factors; the corresponding DNA sequence is called homeobox
Homologous	similar due to common ancestry
Homozygote	the two copies of a given gene are identical
Hybridization	formation of double-stranded RNA or DNA molecules from complementary single strands
Imaginal discs	groups of embryonic cells in the *Drosophila* larva that later form the structures of the adult fly
Implantation	also called nesting or nidation; the anchoring of the blastocyst into the maternal uterine wall
Imprinting	sex-specific determination of the activity of certain genes that are active in early mammalian development
Induction	determination of the fate of cells or tissue by external signals
Inner cell mass	the inner cells of a blastocyst that will later form the embryo as well as extraembryonic membranes
Intron	DNA region of a gene that is transcribed but not translated into protein
In vitro fertilization/IVF	fertilization of mammalian eggs in a culture dish
Germ line	cell groups forming the germ cells or gametes, egg, and sperm
Ligand	signaling molecule binding to a receptor molecule in the cell membrane
Maternal	from the mother
Meiosis	maturation divisions leading to the formation of germ cells
Mesenchyme	tissue consisting of loosely connected cells

Mesoderm	central cell layer of the embryo that later forms the musculature, heart, and gonads, among other structures
Metamorphosis	transformation; development of the fly from the imaginal discs of the larva
Microfilaments	bundles of actin molecules that are part of the cytoskeleton
Microtubules	chains of tubulin molecules that are part of the cytoskeleton
Mitosis	division of the nucleus preceding cell division
Morphogen	substance that has different effects depending on its concentration
Mutation	a change in the structure of a gene
Neural crest	cell groups on the top rim of the neural tube that migrate into the body to form a variety of different structures
Neural tube	folded tube of the ectoderm that later forms the nervous system
Notochord	supporting rod-like organ that stiffens the embryonic axis in chordates
Oncogene	allele of a gene that causes cancer
Organelle	complex structure within a cell that has a certain function
PDG	preimplantation diagnosis; genetic test of individual cells of the embryo in vitro before implantation in the uterine wall
Phenotype	the way genetic activity manifests itself in the organism
Placenta	organ nurturing the mammalian embryo; consists of maternal and embryonic tissue
Placodes	cell groups in the ectoderm of vertebrate embryos that will later form sense organs, feathers, and hair
Plasmid	ring-shaped DNA molecule with few genes
Posterior	at the back end, or rear
Primitive streak	longitudinal indentation through which cells migrate inwardly during gastrulation in the chick and mammalian embryo
Promoter	DNA region of a gene that contains the start point of transcription and the site for binding the RNA polymerase

Receptor	protein molecule in the cell membrane activated by the binding of a ligand
Recombination	exchange of sections between the two homologous chromosomes during meiosis
Replication	DNA doubling
Repressor	transcription factor blocking transcription
Ribosomes	organelles, consisting of protein and RNA, where protein synthesis takes place
RNA	ribonucleic acid; single-strand chain molecule
RNA polymerase	enzyme that transcribes DNA and forms RNA
Signal transduction	signaling a receptor activation to the inside of a cell that leads to a change in genetic activity
Soma	the totality of body cells of an organism
Somites	segments in the mesoderm of vertebrate embryos forming the vertebrae, musculature, and the lower layer of the skin
Stem cells	cells able to renew themselves by division and to produce differentiated cells
Syncytium	multinuclear tissue, or cells with many nuclei
Taxon	group of species originating from a common ancestor
Tracheae	tubes filled with air that provide oxygen to the body of insects
Transgenic	an organism containing an implanted foreign gene in its genome
Transcription factor	a protein binding to the promoter or enhancer of a gene thereby guiding transcription
Transcription	the creation of an RNA copy of DNA
Translation	translation of RNA into protein; protein synthesis with RNA as the matrix
Trophectoderm	outer cell layer of the blastocyst
Tumor suppressor-gene	genes that in their normal form block cell division; in a mutated form, cell division is unhindered
Uterus	organ of the mammalian female harboring the em bryo during development
Ventral	on the belly side
Yolk sac	extraembryonic sac covering the yolk of the chick embryo
Zygote	diploid egg cell developing after fertilization; considered the starting point of an individual's development

INDEX